Qualitative-Quantitative Research Methodology

Qualitative-Quantitative Research Methodology

Exploring the Interactive Continuum

Isadore Newman and Carolyn R. Benz

Southern Illinois University Press
Carbondale and Edwardsville

Library of Congress Cataloging-in-Publication Data

Newman, Isadore.
    Qualitative-quantitative research methodology : exploring the interactive
continuum / Isadore Newman and Carolyn R. Benz.
    p.   cm.
Includes bibliographical references and index.
    1. Research—Methodology. I. Benz, Carolyn R., 1942– II. Title.
Q180.55.M4N49 1998
001.4'2—dc21   97-23117
ISBN 0-8093-2150-5 (pbk. : alk. paper)   CIP

The paper used in this publication meets the minimum requirements of American National Standard for Information Sciences—Permanence of Paper for
Printed Library Materials, ANSI Z39.48-1984. ♾

Dichotomies have their manifest utility, as well as their latent traps. They offer us an heuristic, an analytical scalpel, if you will, by which we can cut phenomena into slices thin enough for us to examine. This, of course, is a useful function, as long as we agree that analyzing the links, the many subtle membranes between the dichotomous end points is a critical legitimate task for any serious analysis. In fact, if dichotomies are at all still useful in a modern world of concatenated complexities it is because the tension between the antithetically conceived end points represents the important possibilities for creativity, ambiguity, paradox, uncertainty, ambivalence, imagination, synthesis, and vision.

—Jean Lipman-Blumen,
"The Creative Tension Between Liberal Arts and Specialization"

Contents

Contents

Figures

Preface

QUALITATIVE AND QUANTITATIVE RESEARCH STRATEGIES and their underlying presuppositions have been increasingly debated since the early 1980s as though one or the other should eventually emerge as superior. We reject the dichotomy assumed by this debate.

We take the position that the two philosophies are neither mutually exclusive (i.e., one need not totally commit to either one or the other) nor interchangeable (i.e., one cannot merge methodologies with no concern for underlying assumptions). Rather, we present them as interactive places on a methodological and philosophical continuum based on the philosophy of science. A researcher tests a theory and, as results feed back to the original hypothesis, both inductive and deductive processes are operational at different points in time; qualitative and quantitative methods are invoked at different points in time; and feedback loops facilitate maximizing the strengths of both methodologies.

We intend this book to meet the general needs of two audiences: research designers and research consumers. First, it is imperative that researchers understand both the determinative nature of the research questions they ask and the assumptions on which they build their designs. This book is intended to assist in building effective designs that are question-based. Second, sophisticated consumers of research need ways to assess the truth value of research findings. The practical approach to criticizing studies will enhance the quality of the judgments research consumers are able to make.

A unique contribution to research practitioners and consumers, the book is founded on the underlying philosophical assumptions of both paradigms and serves mainly as a practical tool. Graphic depictions and narrative descriptions present research as a holistic endeavor; that is, both qualitative and quantitative paradigms coexist in a unified real world of inquiry.

Graduate students and social science faculty have already applied the ideas contained within this book, using earlier drafts and previous presentations. It would be most effectively used as a supplementary book in a graduate-level research-methods course. It could be used by faculty in the behavioral and social sciences to assist in their own research and their work with master's and doctoral students. Education and psychology are our areas of teaching and research, and the ideas are certainly applicable to these fields. However, the ideas and methods are also applicable to the fields of sociology, economics, political science, anthropology, business, and social work. Research workers outside the university will find this a useful supplement to other research manuals.

The purpose of this book is not to teach qualitative and quantitative methods. That is the purpose of other books. Our aim is to have consumers and planners of research think carefully about the consistency between designs and research questions. We assume that the reader has had at least an introductory course in statistics. Fundamental conceptualization of research constructs will help the reader, but a comprehensive, in-depth understanding of neither statistics nor ethnographic strategies is necessary to use the ideas we propose. We have included a glossary to clarify those terms necessary to understand the interactive continuum concept.

We intend that readers be able to analyze their own research plans (and critique others' work) to determine the match between research plans and research methods. Higher-quality research should result from this more thoughtful approach to research methods, an approach that encourages researchers to look beyond narrow dichotomized "either-or" biases toward qualitative and quantitative methods.

Preface

We gratefully acknowledge the assistance of and thoughtful input by many individuals; most important are the students from whom we continually learn and who help us refine our ideas. Among many who have helped us are John Smith, Barbara Moss, Gary Shank, Suzanne MacDonald, Jill Lindsey-North, Ellis Joseph, Susan Tracz, Bill Petrello, Camille Alexander, and Marianne von Meerwall.

Qualitative-Quantitative Research Methodology

1

Qualitative-Quantitative Research: A False Dichotomy

Introduction

BELIEVING THAT THE RESEARCH QUESTION was even more fundamental than the paradigm one felt allegiance to, several years ago we began to discuss the qualitative-quantitative debate from that perspective. The dichotomy and the debate disappeared, and the ideas presented here began to develop.

This book describes our stance at a point in time, not the conclusions of our ideas, which continue to emerge, to grow, and to build from our work as researchers and as teachers. While clearly a work in progress, which continues to evolve, the framework of an interactive continuum presented here has been enlightening to colleagues and students who operate within the current world of often-misunderstood and frequently debated paradigm shifts.

At the conclusion of chapter 1, the reader should be able to

1. Describe the history of qualitative and quantitative research methods and the debate about their relative values

2. Describe the typical purpose and outline of qualitative research

3. Describe the typical purpose and outline of quantitative research

4. Discuss the advantages and disadvantages of a dichotomy versus a continuum conceptualization of research design

Qualitative and quantitative research have philosophical roots in the naturalistic and the positivistic philosophies, respectively. Virtually all qualitative researchers, regardless of their theoretical differences, reflect some sort of individual phenomenological perspective. Most quantitative research approaches, regardless of their theoretical differences, tend to emphasize that there is a common reality on which people can agree.

From a phenomenological perspective, Douglas (1976) and Geertz (1973) believe that multiple realities exist and multiple interpretations are available from different individuals that are all equally valid. Reality is a social construct. If one functions from this perspective, how one conducts a study and what conclusions a researcher draws from a study are considerably different from those of a researcher coming from a quantitative or positivist position, which assumes a common objective reality across individuals. There are different degrees of belief in these sets of assumptions about reality among qualitative and quantitative researchers. For instance, Blumer (1980), a phenomenological researcher who emphasizes subjectivity, does not deny that there is a reality one must attend to.

The debate between qualitative and quantitative researchers is based upon the differences in assumptions about what reality is and whether or not it is measurable. The debate further rests on differences of opinion about how we can best understand what we "know," whether through objective or subjective methods.

William Firestone (1987), in an article in the *Educational Researcher*, differentiates qualitative from quantitative research based on four dimensions: assumptions, purpose, approach, and research role. Regarding assumptions, Firestone asks: is objective reality sought through facts or is reality socially constructed? Related to purpose, he asks: is it looking for causes or for understanding? To determine approach, he asks whether the research is experimental/ correlational or a form of ethnography. Lastly, related to the researcher's role, he asks whether the researcher is detached or immersed in the setting.

Shaker (1990), in a discussion of program evaluation models, presents them as a metaphorical journey—moving from quantitative perspectives in the past to more recent naturalistic and qualitative assumptions. While positing a chronological continuum, Shaker would not seem to oppose our notion of question-driven research and evaluation. While he describes the "new identity" for evaluation as being "based on naturalistic approaches," he places this in the context of a "pragmatic commitment to finding methods that yield results in practice as we find it, rather than as we wish it to be" (p. 355).

The qualitative, naturalistic approach is used when observing and interpreting reality with the aim of developing a theory that will explain what was experienced. The quantitative approach is used when one begins with a theory (or hypothesis) and tests for confirmation or disconfirmation of that hypothesis.

It is important here to set the stage for abandoning the dichotomy. To do so, we examine a few of the key events in the chronicle of scientific evolution that established the debate in the first place. As long as one view of how we can explain the workings of the world reigns supreme, there is no debate. The debate rests on a dichotomy characterized by a lessening of the dominance of one paradigm over another, leveling the playing field so that the debate could occur. In fact, the debate may be but one more phase in the ebb and flow of an ever-changing philosophy of knowledge. For example, in *The Enlightened Eye*, Eisner (1991) cautions against the dichotomy and asserts that qualitative and quantitative research can be combined. He warns against qualitative researchers merely adopting a "soft form of positivism" (p. 167).

The genesis of the current qualitative-quantitative debate in educational research occurred as far back as 1844, when Auguste Comte claimed that the methods of natural science could be justified in studying social science (1974; see also Vidich & Lyman, 1994). Science, in this view, is the collection and study of facts that can be observed through sensory input. These are the traditional data investigated by natural scientists—the physicists, the chemists, the biologists. This view holds that *true* science is accumulated through the study of phenomena that can be physically sensed, observed, and counted. The "unknowables," as Herbert Spencer described them in his 1910 essay, those things that cannot be sensed but might rely on reason or thought, are banished from scientific investigation. Both Comte and Spencer were positivists.

Interestingly, this "positivism" was a move away from a more speculative, more "unknowable" view. It was a move away from relying on theological and metaphysical explanations of the world. It was a move toward what could be "positively" (confirmed through sensory data) determined. The philosophy maintained a grip on social science from the late 1800s through the early 1900s.

In the early 1900s, John Dewey, among others, questioned the absolutism of this position, viewing science as not separate and distinct from problem solving. His pragmatism considered science less rigidly than did the positivists. In his *Sources of a Science of Education* (1929), written some time after his initial speculations, he pointed out that practice should be the ground of our inquiry. Because of the value placed on experience for learning and the emphasis on practice, he appreciated the deeper complexity of what educational and social scientists study. During the same period, a group of scholars who made up what became known as the famous Vienna Circle met and developed a new philosophy of science, logical positivism. Supporting Comte's positivism, they combined it with the symbolic logic of mathematics. Hypotheses derived using the rigor of mathematics (the symbolic) could be combined with fact gathering (the positivism) to test their confirmability (which was eventually modified to *disconfirmability*). Although counter to an impetus by Dewey to diffuse the positivistic as-

4

sumptions made by researchers, this hypothetico-deductive system was dominant in the middle years of the 20th century in psychology and sociology. Education, which borrowed traditions of inquiry from these disciplines, was affected as well. The respect for precision in measurement, mathematically systematic tests of hypotheses, and a quest for value-free science solidified this paradigm.

During the 1940s and 1950s, the quantitative paradigm dominated the social science and the educational research scene. Behaviorists and organizational theorists utilized empirical fact gathering and hypothesis testing almost exclusively in studying educational and social phenomena.

In the mid-1960s, while the quantitative perspective continued to prevail, a shift began as skepticism toward the domination of logical positivism and the evident chasm between human social systems and mathematical logic grew. New epistemologies began to emerge that acknowledged, for example, the value-laden nature of human social interactions. That human beings construct reality for themselves and that knowledge itself is transmitted in social ways were beginning to be assumed. Questions arose about the tenability of applying natural science methodology to these complex human dynamics.

In 1962, in *The Structure of Scientific Revolutions*, the most significant work on this issue, Thomas Kuhn explored the shifts in science's dominant paradigms. His doctorate in theoretical physics led him to look back into the history of science as he sought to know more about its foundations. He describes how, by randomly exploring the literature, he was exposed to Jean Piaget and, in the late 1950s, to a historical analysis of social science and psychology. Kuhn's study of methodology drove him to leave physics and become a historian of science. He conceptualizes the notion of paradigms, "universally recognized scientific achievements that for a time provide model problems and solutions to a community of practitioners," (1970, p. viii) and proposes that competing paradigms emerge chronologically when the dominant one no longer serves the explanatory needs of the scientific community. For the most part, using the context of physics from the perspectives of Sir Isaac Newton and Albert Einstein, Kuhn explains these periods of competition, or scientific revolutions, in the natural sciences. He acknowledges that competing paradigms can possibly co-exist on equal footing following such a revolution, or "paradigm shift," although, he cautions, it may be only rarely possible.[1] He proposes that the predominant paradigm affects researchers not only methodologically but also in how they see the world. Kuhn's conceptualization of "paradigm" has been reinterpreted by others since his work, and many definitions are incorporated in the literature of the 1990s.

The quantitative paradigm continued to reign over social science and, according to Culbertson (1988), prevailed in education until the mid-1980s. At that time the logical positivists were losing supremacy. (The strong traditional bias toward quantitative science might even be based on Americans' preference for facts we can observe and count, a sense that that's what science "is.")

Concurrent with Kuhn's early notions of paradigms in the 1960s, society was undergoing radical changes. While some began to question the efficacy of the positivists' tools in explaining human organizational and social phenomena, education was moving into a more complex social context. Culbertson points to such 1960s and 1970s issues as racial integration, poverty, equal opportunity, schools as tools in global economic competition, the Soviet Union's threat to our math and science preeminence, and the need to account for the success and failure of the nation's children and posits that, in this context of increased complexity, some began to search for policy tools that the quantitative paradigm did not seem sufficiently able to explain. That education served economic, political, and policy ends enhanced the opportunity for scholars interested in the culture of schools to begin to use anthropological strategies in their inquiry. These same interests fed the scholars' attempts to approach their research from the perspective of the critical theorists, as well as that of the feminists. Although always an important issue, the policy makers' interest in the world of classroom practice grew, and they increasingly expressed concerns that research and practice were unconnected and that this disconnection was in part due to the use of tightly controlled laboratory-like quantitative assumptions. A move among some social scientists in the direction of deriving theory *from* practice, rather than the other way around, characterized this change as well.

Graduate programs preparing educational and social science researchers as well as professional journals have increasingly directed their attention toward qualitative research. Allotting time and space to what had been considered the "alternative" paradigm led to wide discussions in the journals and at professional meetings. The editors of the *American Educational Research Journal*, for example, announced in 1987 that particular emphasis on qualitative methodology would be forthcoming as they evaluated manuscripts. This was a major legitimation of the paradigm for educational researchers. A plethora of books, articles, and presentations on the trustworthiness of the qualitative paradigm materialized. Some extolled the virtues of qualitative research as the only avenue to "truth," while others claimed that only by holding onto the quantitative traditions can we have confidence in our knowledge base. In many forums the debate was manifest. Which is more scientific: the deductive methods of the logical positivists (quantitative researchers) or the inductive methods of the naturalists (qualitative researchers)? Can the results of qualitative research be generalized as are the results of quantitative research? Can science be value laden (qualitative) or only legitimate if value free (quantitative)? What epistemological assumptions are violated by adopting one paradigm or the other?

While to some the debate has ended, to others, especially those we encounter in researcher-preparation programs, the debate has either not yet materialized to the full extent of its fury or continues unabated. Our strong sense is twofold. First, we continue to prepare students for an "either-or" world, a dichotomous world, that no longer exists. We still prepare students who leave our colleges and universities with a monolithic perspective. Either they become well-trained statisticians, or they become cultural anthropologists, methodologically weak in asking research questions and in justifying either one or the other set of strategies. Second, researchers in education and in the social sciences have not yet constructed a way to ensure their success in utilizing both paradigms. The interactive continuum model in this book serves as a kind of framework directed toward both those needs.

The dichotomy of qualitative and quantitative research is one we deny but one we exploit here for heuristic purposes. The dichotomy, while not an ontological construct, does allow us to separate the idea. We slice it thin to examine it and make the case in this chapter that it does not exist in the scientific research realm.

In chapter 2, we elaborate on the notion of the interactive continuum. We discuss the construct of validity, review methods, and address the strengths and weaknesses of both paradigms in chapter 3. In chapter 4, we discuss strategies to increase validity in quantitative and qualitative methods.

Chapter 5 contains approaches to applying the continuum by asking questions to assess whether the research purpose is consistent with the assumptions and methods of that research. We present applications of the model to four articles from education and counseling. In the final chapter, chapter 6, we summarize the interactive continuum, its application, and how its use can enhance educational research by clarifying a *unified* philosophy of science to the novice, as well as by expanding the perspectives of the experienced researcher. We make the case that, rather than there being a dichotomy between qualitative and quantitative approaches, research is based on a unified philosophy of science and can be more appropriately conceptualized as an interactive continuum. This approach can be transformed into an operational model to assist both in critiquing published research and in planning one's own research.

All research in education stands on basic underlying assumptions.

This is true for quantitative methods as well as qualitative methods. To the extent that these assumptions withstand the scrutiny of scientific inquiry, the methods can be supported, taught to novice researchers, and used professionally and ethically without reservation. Since the mid-1980s when quality in all educational professions came under public review, it has become particularly crucial to delineate the foundational bases of educational research. Within the realm of this book, such bases will be examined.

Qualitative Versus Quantitative: A False Dichotomy

All behavioral research is made up of a combination of qualitative and quantitative constructs. In this book, the notion of the qualitative-quantitative research continuum, as opposed to a dichotomy, is explored on scientific grounds. We believe that conceptualizing the dichotomy (using separate and distinct categories of *qualitative* and *quantitative* research) is not consistent with a coherent philosophy of science and, further, that the notion of a continuum is the only construct that fits what we know in a scientific sense. A secondary theme is equally important; that is, what are known as qualitative methods are frequently beginning points, foundational strategies, which often are followed by quantitative methodologies.

Qualitative research methods are those generally subsumed under the heading *ethnography*. Other headings and names include *case studies, field studies, grounded theory, document studies, naturalistic inquiry, observational studies, interview studies,* and *descriptive studies.* Qualitative research designs in the social sciences stem from traditions in anthropology and sociology, where the philosophy emphasizes the phenomenological basis of a study, the elaborate description of the "meaning" of phenomena for the people or culture under examination. This is referred to as the *verstehen approach.* Often in a qualitative design only one subject, one case, or one unit is the focus

9

of investigation over an extended period of time. According to Glaser and Strauss (1967), qualitative data are often coded a posteriori from interpretations of those data.

Quantitative research, on the other hand, falls under the category of *empirical studies*, according to some, or *statistical studies*, according to others. These designs include the more traditional ways in which psychology and behavioral science have carried out investigations. Quantitative modes have been the dominant methods of research in social science. Quantitative designs include experimental studies, quasi-experimental studies, pretest-postest designs, and others (Campbell & Stanley, 1963), where control of variables, randomization, and valid and reliable measures are required and where generalizability from the sample to the population is the aim. Data in quantitative studies are coded according to a priori operational and standardized definitions.[2]

It is necessary to adopt some standard by which one can measure whether the qualitative, the quantitative, or a continuum that includes both methodologies is the most effective mode in reaching truth. We assume the standard of science as a way of knowing.

Mouly (1970) asserts that, although there are two ways other than science to "know" something (i.e., "experience" and "reasoning"), only through science can we generalize and provide for theory building. Some would have us believe that we can know something based on "authority." This basis has similarly been discredited because of the frequent inability to verify the facts, as well as the conflicting points of view among authorities. Other philosophers (described in McAshan, 1963) go even further and suggest one can "know truth" also through "serendipity," "intuition," "compromise," and "consensus." Conjecture surrounding how we can know about truth, repeatable and verifiable truth, runs the gamut from "faith" to simple sensory perception. The assumption here is that science, as reflected in the scientific method, is the only defensible way of locating and verifying truth. Therefore, the criteria for comparison of the constructs underlying the dichotomy (qualitative vs. quantitative) and the interactive continuum (qualitative to quantitative to qualitative, etc.) are their scientific bases

The search for knowledge (or "truth") is the purpose of research.[3] This search and, concomitantly, this research is most effective when built on the scientific method. In the ongoing debate between the positivists and the naturalists we tend to support the idea that the modern-day scientific method is both inductive and deductive, objective and subjective. Design validity is more likely to be built into studies when the researcher is open to both paradigms rather than precluding one or the other. When faced with the question, "Which is better?" we would refuse to answer; indeed, we would be *unable* to answer, given the choices presented. There is no such answer. The better paradigm (qualitative or quantitative) is the one that serves to answer the specific research question.

We began our thinking on these issues over a decade ago. Our thoughts began to solidify in an interactive continuum model in 1985. Others have written about integrating qualitative and quantitative methods. Cook and Reichardt (1979) predates our original work and, like us, they suggest that the researcher's method can be separated from the researcher's worldview. Their book differs from ours in that their ideas are presented in an introductory essay to a collection of essays by research methodologists. Their purpose was to bring together the combined works of many who were then struggling with the issues. Michael Patton (1980) presents a diagram of what he calls "mixed paradigms" in his book, *Qualitative Evaluation Methods*. His conceptualization, like ours, acknowledges that, between the qualitative and quantitative paradigms, there is a continuum of methods. His book, however, addresses qualitative methods only. It is not an exhaustive examination of assumptions, methods of research, and ways to critique research studies as we intend ours to be.

Creswell (1994), too, has authored a volume, *Research Design: Qualitative and Quantitative Approaches*, and he intends it to assist the researcher in making decisions about design. His book seems most closely focused on writing a dissertation proposal, and it is organized in that sequence. It does not include critiquing research as ours does, and he does not present an overall model of his thinking. The book is replete with examples from both qualitative and quantitative studies. Our book contributes to the current discourse on research methods and assumptions underlying social science research by

1. Depicting an overall model of qualitative-quantitative interactive continuum

2. Suggesting ways to assess quality of published research

3. Providing a strong emphasis on validity

In the last decade, a debate has continually raged as though one or the other paradigm should eventually win. Discounting the debate is not the issue of importance. The key issue, we believe, should be improving the quality of research through an integrated way of viewing qualitative and quantitative research methods. Both paradigms coexist in the world of inquiry, and together they form an interactive continuum. Operationalizing this model is the focus of the rest of our book.

2

Qualitative and Quantitative Research Methods: An Interactive Continuum

TYPICALLY, ANY DISCUSSION OF RESEARCH METHOD is dichotomized and presented in either a quantitative or a qualitative category because the two paradigms have been assumed to be polar opposites and, among some, even separate and distinct scientific absolutes. These claims and counterclaims have been the genesis of the debate described in chapter 1. We conceptualize science more broadly than either of these opposites implies.

Science fundamentally requires a set of systematic rules of procedure. We would not adhere only to philosophers of science who, like Karl Popper in his earlier views, claim that only those hypotheses that can lead to claims of falsifiability are "scientific" (1962; see also Hempel, 1965). Despite his contribution that science includes more than verifiability (i.e., falsifiability), those two issues alone are not sufficient. If they were, that view would exclude the metaphysical, the speculative, the existential, and the heuristic. Diesing (1991) has so rightly claimed that it would be better to admit all kinds of statements, both verifiable and falsifiable, into the realm of potential scientific investigation. We would go further, as a matter of fact, and include the premise of the naturalists: that science includes interpreting the constructed reality one experiences. This assumption is most associated with qualitative research. As a picture of "lived reality," that knowledge, too, can be examined in scientific ways. Science, once again, embodies a set of rules of procedure. We assume science more broadly than either the traditional logical empiricists or the naturalists alone define it.

In this chapter, we present a conceptualization of research methods as existing on an interactive continuum rather than as a dichotomy and include discussion of scientific inquiry, the purpose of research, the kinds of questions that are typically posed, and our fundamental assumption that each question dictates the research method. We argue that both qualitative and quantitative strategies are almost always involved to at least some degree in every research study.

At the conclusion of this chapter, the reader should be able to

1. Explain how "science" incorporates all ways of knowing across the research paradigms

2. Describe the qualitative approach to research

3. Describe the quantitative approach to research

4. Justify the central place of theory in qualitative and quantitative methods

5. Describe the interactive continuum

From Mouly (1970) to McMillan and James (1992), the traditional assertion made in research texts has been that, although there are ways other than science to "know" things (i.e., experience and reasoning), only through science can we generalize and provide for theory building and testing. Our assumption is that science and the traditions science has maintained in its rules of procedure are fundamental requirements for locating and verifying knowledge.

Because the scientific process and its rules allow us to acquire knowledge, we can, assume no singular epistemology. Likewise, we assume that no one method to acquire knowledge is superior. That there are clear a priori assumptions and rules of procedure consistent with those assumptions becomes the standard of science. One, then, can determine whether the qualitative, the quantitative, or a continuum including both methodologies is most effective.

14

The researcher must begin with the nature of the research question. According to our assumptions of science, the research question must be considered first, the accessibility of the data second, and whether the data are or are not quantified, according to the design of the study, third. In other words, the decision about what data to collect, as well as what to do with those data after they are collected, should be dictated by the research question.

Given this standard for science as we present it, the dichotomy no longer exists. The paradigm of methods is inclusive and follows naturally from the research question. For example, Miller and Lieberman (1988) observe in education a "new synthesis." In their review of studies of school improvement, they acknowledge the different sets of assumptions underlying qualitative and quantitative studies but describe studies that combine the "technological" perspective of the quantitative with the "cultural" perspective of the qualitative.

Methods of systematic scientific inquiry can apply to all areas of inquiry, physical as well as behavioral. As noted by Keppel (1973), for example, the scientific method can show two bits of behavior that tend to appear together in nature and then use this fact to predict the occurrence of one behavior from the occurrence of the other. While science traditionally is linked in most people's minds to the world of physics, medicine, chemistry, and other hard sciences, Mouly (1970) claims otherwise. Social phenomena, he maintains, can be examined by scientific methods just as physical phenomena are. It is only the inadequacy of our current knowledge that limits our ability to predict something as complex as human behavior.

Along with other behavioral researchers, we tend to agree with Mouly's assertion. The paradigm of positivism (quantitative research) continues to dominate social and behavioral science. It is steeped in historical tradition. For one thing, the training of research methodologists in social science and education has been heavily weighted on the side of quantitative research designs and statistics. The challenge from qualitative adherents over the past 30 years has not been successful in overthrowing that dominance but has led to the debates between the advocates of quantitative research and the advocates of qualitative research.

Instead of an us-them dichotomy, however, we take the broader view that the scientific tradition is being enhanced. Science is both positivistic and naturalistic in its assumptions. Two fundamental epistemological requirements are made of the researcher: one must clearly and openly acknowledge one's assumptions about what counts as knowledge and maintain consistency in those assumptions and the methods that derive from them. To us, this is what makes the research scientific.

Qualitative Methods Conceptualized

In their *Handbook of Qualitative Research*, Denzin and Lincoln (1994) acknowledge that qualitative research means different things to different people. They offer what they call a "generic definition."

> Qualitative research is multimethod in focus, involving an interpretive, naturalistic approach to its subject matter. This means that qualitative researchers study things in their natural settings, attempting to make sense of, or interpret, phenomena in terms of the meanings people bring to them. Qualitative research involves the studied use and collection of a variety of empirical materials—case study, personal experience, introspective, life story, interview, observational, historical, interactions, and visual texts —the described routine and problematic moments and meanings in individuals' lives. (p. 2)

Qualitative data are defined by Patton (1990) as "detailed descriptions of situations, events, people, interactions, observed behaviors, direct quotations from people about their experiences, attitudes, beliefs, and thoughts and excerpts or entire passages from documents, correspondence, records, and case histories" (p. 22). Theory's place in qualitative methods is quite different from that in quantitative methods. Grounded theory methodologists, one group of qualitative investigators, are examples of theory builders: theory emerges from the data and is thus grounded in the data rather than being abstract or tentative (Glaser & Strauss, 1967). Compared to the hypothesis-testing method, grounded theory is instead developed by (1) entering the fieldwork phase without a hypothesis; (2) describing what happens; and (3), on the basis of observation, formulating explanations about why it happens (Lincoln & Guba, 1985; Patton, 1990). Instead of coming from the conceptual level to the empirical level, they begin at the empirical level (data collection) and end at the conceptual level.

Sherman and Webb's (1988) edited volume includes a variety of qualitative methods in individual essays by a number of researchers. Marshall and Rossman (1989), Glesne and Peshkin (1992), and others also describe ways to design qualitative research.

Inductive reasoning and deductive reasoning are both subsumed under scientific inquiry, yet they characterize a distinction between purely qualitative and purely quantitative methods. Patton (1990), in fact, states the separation even more strongly when he remarks: "The cardinal principle of qualitative analysis is that causal relationships and theoretical statements be clearly emergent from and grounded in the phenomena studied. The theory emerges from the data; it is not imposed on the data" (p. 278). We do, however, take exception to the latter idea. Theory does not emerge independent of the person interpreting the data. Data do not develop theory; people do.

While we are comfortable living within the realm of the qualitative-quantitative interactive continuum as a system of our beliefs about research, our position within the wider academic world of educational research probably resonates most with Norman Denzin's notion of postpositivism (1994). Denzin describes the four responses that have been made to the legitimation crisis. First, the positivists apply the same four criteria to qualitative research as to quantitative research: "internal validity, external validity, reliability and objectivity" (p. 297). Second, the postpositivists believe a separate set of criteria needs to be developed for qualitative research. Denzin characterizes those who fall in this group as often creating a set of criteria that parallels that of the positivists but is adjusted to naturalistic research. Third, the postmodernists claim that there can be no criteria for judging qualitative research. Fourth, according to Denzin, the critical poststructuralists believe that new criteria, completely different from those of both the positivists and the postpositivists, need to be developed. It is with this last group, the critical poststructuralists, that Denzin aligns himself.

Within Denzin's structure, our position aligns with his second category, postpositivism, because we believe a different set of criteria should be applied to assess qualitative research. The criteria established by Lincoln and Guba (1985) differ from those established for quantitative research, but they are philosophically derived from them.

Denzin (1994) describes the "legitimation crisis" as the concern for the validity of qualitative research. The postpositivists call for a set of rules of procedures to establish validity. According to Denzin:

> A text's authority, for the postpositivist, is established through recourse to a set of rules that refer to a reality outside the texts. These rules reference knowledge, its production and representation. . . . Without validity (authority) there is no truth, and without truth there can be no trust in a text's claim to validity (legitimation). (p. 296)

Quantitative Methods Conceptualized

Quantitative research is frequently referred to as *hypothesis-testing research* (Kerlinger, 1964). Typical of this tradition is the following common pattern of research operations in investigating, for example, the effects of a treatment or an intervention. Characteristically, studies begin with statements of theory from which research hypotheses are derived. Then an experimental design is established in which the variables in question (the dependent variables) are measured while controlling for the effects of selected independent variables. That the subjects included in the study are selected at random is desirable to reduce error and to cancel bias. The sample of subjects is drawn to reflect the population. After the pretest measures are taken, the treatment conducted, and posttest measures taken, a statistical analysis reveals findings about the treatment's effects. To support repeatability of the findings, one experiment usually is conducted and statistical techniques are used to determine the probability of the same differences occurring over and over again. These tests of statistical significance result in findings that confirm or counter the original hypothesis. Theory revision or enhancement follows. This would be a true experiment.

These procedures are deductive in nature, contributing to the scientific knowledge base by theory testing. This is the nature of quantitative methodology. Because true experimental designs require tightly controlled conditions, the richness and depth of meaning for participants may be sacrificed. As a validity concern, this may be a limitation of quantitative designs.

The Interactive Continuum

When considering methods from both ends of the continuum and their scientific base (their basis in what we call *knowing repeatable facts*), different assumptions are apparent. The concept of a continuum is a more comprehensive approach. Evidence of such a continuum is demonstrated by an increasing number of researchers who apply multiple methods to their research and by the increased popularity of multimethod approaches in sociological research.[1] Despite the debate, these ideas are not new. They are now more strongly emphasized. More than 25 years ago, Mouly (1970) alluded to multiple-perspective research as the essence of the modern scientific method. . . . Although, in practice, the process involves a back-and-forth motion from induction to deduction, in its simplest form, it consists of working inductively from experience to hypotheses, which are elaborated deductively from implications on the basis of which they can be tested. (p. 31)

If we accept the premise that scientific knowledge is based upon verification methods, the contributions of information derived from a qualitative (inductive) or quantitative (deductive) perspective can be assessed. It then becomes clear how each approach adds to our body of knowledge by building on the information derived from the other approach. This is the premise of the interactive continuum. A schema that (philosophically) depicts this continuum appears in Figure 1.

The place of theory in both philosophies is shown to overlap. This is where the concept of the continuum is most clear. For the qualitative researcher, the motivating purpose is *theory building*; while for the quantitative researcher, the intent is *theory testing*. Neither the qualitative research philosophy nor the quantitative research philosophy encompasses the whole of research. Both are needed to conceptualize research holistically.

In general, the qualitative researcher follows the sequence shown in the circles (labeled with letters). At circle *A*, data are collected, interpreted, absorbed, and experienced. At circle *B*, the data are analyzed; and at circle *C*, conclusions are drawn. From those conclusions, hypotheses are created (circle *D*). Those hypotheses can be used to develop theory (circle *E*), the goal of the research question.

Quantitative research begins with theory (square 1). From theory, prior research is reviewed (square 2); and from the theoretical frameworks, hypotheses are generated (square 3). These hypotheses lead to data collection and the strategy needed to test them (square 4). The data are analyzed according to the hypotheses (square 5), and con-

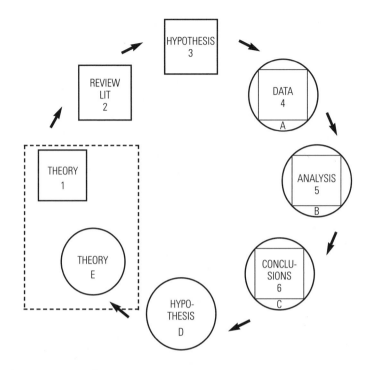

Conceptually, in *our* model, the "theory" is neither at the beginning nor at the end—but the square and circle would overlap and continue the cycle closing the qualitative-quantitative gap. Neither the squares (quantitative) nor the circles (qualitative) make a complete whole.

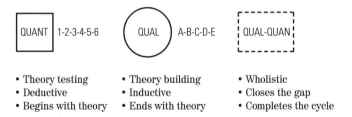

• Theory testing	• Theory building	• Wholistic
• Deductive	• Inductive	• Closes the gap
• Begins with theory	• Ends with theory	• Completes the cycle

Figure 1. The qualitative-quantitative philosophy of educational research methodology conceptualized

clusions are drawn (square 6). These conclusions confirm or conflict with the theory (square 1), thereby completing the cycle.

The qualitative-quantitative continuum is strengthened scientifically by its self-correcting feedback loops. In each and every research study, the continuum operates. When one conceptualizes research this way and uses the built-in feedback mechanism, positive things happen that are less likely to occur in a strictly qualitative or a strictly quantitative study. For example, data can be more parsimoniously collected in a quantitative study if the research question has been defined by preliminary document study, participant observation, historical review, or interview. These qualitative foundations of a study enhance its validity. These empirical materials may feed into the data-collection instruments or to the sample selected, altering these components, correcting them for further study.

While there is probably no single representation or schematic diagram that can easily explain the concept of the qualitative-quantitative interactive continuum, Figure 2 explains the model conceptually and summarizes the interrelationships between qualitative and quantitative methods as approaches to scientific inquiry. It is important that the reader understand that this is a simplification of a concept that has an infinite number of combinations.

All research endeavors probably start out with a topic of interest. If pushed, researchers can speculate why this topic is of interest and is of value. Sometimes this speculation becomes structured and formal and takes on the qualities of a theory. However, it can remain loose and informal, based on phenomenological experiences and assumptions. Generally, once the speculation stage is reached, the next step, in both qualitative and quantitative research, is to do a review of the literature. However, there are certain qualitative researchers who believe that one should not enter the research with preconceived notions, that the data should be free from the bias of the researcher's prior knowledge and expectations. Two examples from the literature demonstrate this view.

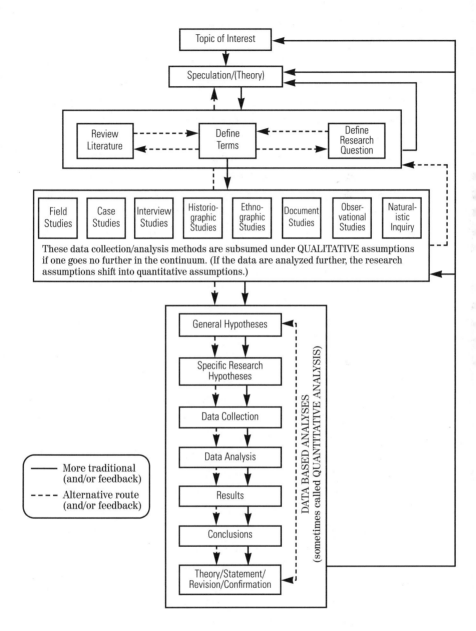

Figure 2. Qualitative-quantitative interactive continuum

Frederick Erickson (1973, cited in Goetz & LeCompte, 1984, p. 2) describes one group of advocates for ethnographic studies who enter the field purposefully assuming a naïveté, while others merely suspend their preconceptions. L. M. Smith (1967) describes how one assumes ignorance in terms of the foreshadowed problem. The question keeps the researcher on the track of the most cogent data. While one is in the field, the research question guides what one attends to; this strategy has become common for qualitative researchers. We see this concept of foreshadowing as not entirely different from the notion of *working hypotheses* among empiricists, defined as those relational statements derived from descriptive research, theory, or personal experiences (see Ary et al.,1980; Rosenthal & Rosnow, 1991).

We believe that one always has preexpectations and that it is important for researchers to be aware of what biases they have. Only through awareness can one control for bias in the data-collection stage. This is the rationale for the schematic structure presented here. At the same time, the reader must be flexible and understand that this diagram is an attempt to represent conceptually the qualitative *and* the quantitative strategies within systematic scientific inquiry. The decision about method rests on the purpose and the assumptions of the research question, which guides the research methods *not* vice versa. The method should not dictate whether the research is qualitative or quantitative; one should not interpret Figure 2 as implying that it does.

The review of the literature can be related directly to the topic, to the background of the topic, or to the applications and usefulness of the topic. Often the literature review, definitions of terms, and the research questions are interdependent. One is an outgrowth of the others or, depending on how much information the researcher has at the beginning, tends to change the others. This interdependence is represented by dotted lines going back and forth between these three elements and by the relationship between this box and the one labeled "Speculation/Theory."

The next section of Figure 2 contains qualitative methods. It is difficult to represent these methods accurately as discrete entities because overlap almost always occurs. One study strategy may use another study strategy within its framework, as well as within its data-collection procedures. For example, if an investigator uses an ethnographic strategy, the collected information might be coded numerically and analyzed statistically in a hypothesis. However, an underlying assumption of the ethnographic method is that one cannot generalize; the researcher cannot begin with a purpose toward generalizability of findings and then carry out the research methods in ways that disallow generalizability.

There is a dotted line that seems to go through the qualitative strategies box into the quantitative methods box. This line represents the fact that, in quantitative research, it appears that one goes from reviewing and defining directly to developing hypotheses and collecting data. In quantitative analysis, this is called the *derivation of hypotheses*. These derivations may be considered qualitative analyses in simplified form. The researcher examines the literature and, based upon this process, he or she derives theoretical expectations, which become the derived hypotheses. The solid line going from the "Review Literature, Define Terms, Define Research Question" box to the qualitative strategies box and its feedback loops are what some individuals will identify as qualitative analysis in its entirety. Other researchers would suggest that one go from that feedback to the quantitative methods box and use it before "appropriate and scientific" conclusions could or should be made from qualitative data. As one can see, the qualitative analysis with its feedback loops can easily modify the types of research questions that will be asked in quantitative analysis research; and the quantitative analysis results and its feedback can change what will be asked qualitatively. Therefore, this model is not only a continuum from qualitative to quantitative but interactive.

An example of the implementation of this approach was presented by Benz and Newman (1986) whose research on teacher preparation was built on this qualitative-quantitative continuum model. Student teachers' quantitative ratings of their experiences on questionnaires followed interpretive analyses of their narrative responses. While students rated one seminar at a low level numerically, it was not until telephone interviews were conducted that it was revealed that the scheduled time the seminar met was most disturbing to the students not the content.

One needs to identify qualitative and/or quantitative research according to the type of question being asked and the type of data being collected. If the data cannot be quantified or are not quantified, then the research is qualitative. If one wishes to terminate the discourse in the scientific process within the qualitative analysis box of this schema, then the research is qualitative. One goes no further in the diagram. If one utilizes the strategies in the quantitative analysis sequence, the research is quantitative.

In the diagram, one can see the feedback loops that facilitate theory revision, see where theory fits in both methods, and, to some extent, understand why theory is never proven absolutely. It is always subject to modifications as new data enter the system. This approach fits and is applicable to both qualitative and quantitative research conceptualizations. Examples of research critiques presented in chapter 5 demonstrate how one study could and should lead to other investigations.

As stated, in the last 15 years or so, proponents of both approaches have assumed that one or the other paradigm would eventually "win."[2] We emphasize that the real issue is improving the quality of research. The focus of the rest of this book is the application of the continuum model in concrete ways to help researchers conduct their own and evaluate their own and others' research.

3

Validity and Legitimation of Research

RESEARCH OUTCOMES ARE OF NO VALUE if the methods from which they are derived have no legitimacy. The methods must justify our confidence. Those who read and rely on research outcomes must be satisfied that the studies are valid, that they lead to truthful outcomes.

In this chapter, we first present a model to show how we connect research questions and methods to the truth value of the outcomes. Next we discuss validity issues in quantitative research through a review of internal- and external-validity criteria. Then we discuss issues of validity in qualitative research, a much more amorphous topic. Contrasted with the classic criteria for the validity of quantitative research described by Campbell and Stanley (1963), the criteria of validity (legitimation) of qualitative research has no consensus. We briefly describe the ideas of Martin Hammersley, Margaret LeCompte, Judith Goetz, Steinar Kvale, Patti Lather, and William Tierney so that we may locate our own views as being closest to Lincoln and Guba's (1985) postpositivist perspective. Last, we list the 13 criteria for truth value that will later become our criteria for designing and critiquing qualitative studies.

At the end of this chapter, the reader should be able to

1. Define validity and legitimation
2. Define research design in both qualitative and quantitative paradigms

3. Differentiate design validity and measurement validity
4. Define internal and external validity
5. Describe the threats to internal validity
6. Describe the threats to external validity
7. Describe ex post facto research
8. Compare and contrast various qualitative research methodologists' conceptualizations of validity
9. Describe the 13 methods that add to the legitimation of qualitative research

The notion of *validity* has a long history and strong consensus among most traditional education researchers. The concept is applied in at least two contexts—in research design (internal and external validity) and in measurement (the validity of the measurement). We struggled to conceptualize validity in these two contexts in our thinking and our model. We ultimately placed the validity of measurement within the category of Campbell and Stanley's "instrumentation" (1963, p. 5). When there is a strong estimate of face validity, content validity, construct validity of one's measurement, for example, the threats to the internal validity of one's design are lessened.

Harry Wolcott has conceptualized the same two validity concepts, but he does so in a chronological way. In his essay in Eisner and Peshkin's 1990 edited volume, *Qualitative Inquiry in Education*, Wolcott claims to be unconvinced that validity needs quite so much emphasis in naturalistic studies (1990a). He suggests that the word *understanding* replace *validity* in qualitative research. Interestingly, he chronicles the evolution of the emphasis on validity over the past three decades as a developmental pattern.

- Test validity
- Validity of test data

- Validity of test and measurement data
- Validity of research data on tests and measurements
- Validity of research data
- Validity of research

He seems to indicate that research validity as a concern grew out of original concerns for measurement validity. Our concern has been validity: the truth value of research outcomes is stronger when both the data and the design are valid. We have deliberately used the words *validity* and *legitimation* in the title of this chapter. Validity has traditionally meant an estimate of the extent to which the data measure (or the design measures) what is intended to be measured. But not all researchers conceptualize the links between design and truth value in this way. *Validity*, as we have defined it in our experience, does not fit all researchers' definitions. *Legitimation* is a recent term that we borrow here to relate to a broader notion of truth value. In essence, it can be considered somewhat parallel with our notion of validity.

To Denzin (1994), for example, truth value becomes almost circular. Without validity, he says, there is no truth, and without truth, there is no claim of validity. He has described this current period in qualitative research as a "crisis of legitimation" (p. 295). Legitimation means that the research methods are consistent with the philosophical underpinnings of the question. For example, the positivist assumes an objective reality; the postmodernist assumes no objective reality and no objective truth (see Appendix A). To a certain extent, the notion of legitimation mirrors our set of consistency questions in chapter 5.

While a perfectly accurate portrayal of our notions of validity across the continuum is not possible, we can outline the major dimensions of our thinking. As much as a diagram is able, Figure 3 depicts our reasoning about connecting questions, methods, and truth value.

29

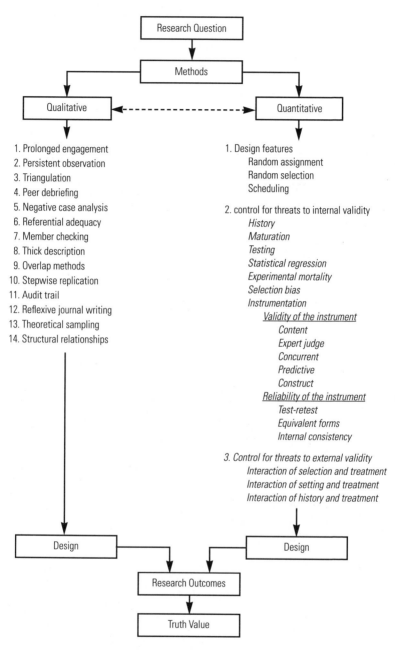

Figure 3. Links between research questions and truth value

Connecting Research Questions, Methods, and Truth Value

Figure 3 should not be interpreted to mean that the decisions embedded within it are necessarily made in this linear and hierarchical fashion at each point in the process. On the other hand, the research question *always* initiates any set of decisions the researcher makes. Within the dimensions following the question, however, the decisions may be cyclical. (Quantitative and qualitative methods may cross over in particular situations, thus the dotted line connecting them in Figure 3). What is not negotiable is the overall linear connection between question, methods, and truth value. While it is necessary to accept that design decisions are "emergent" in good qualitative research, the researcher's thinking is not emergent. In the researcher's thinking, the conceptual linking of question–methods–truth value must be linked, defensible, and predetermined.

As mentioned before, the research question guides what methods we select. For that reason, the figure begins at the top with the "Research Question" box which leads to the next construct—"Methods." Because our overall intent is to present research holistically, not as a dichotomy, we wanted to discuss both paradigms in similar conceptual ways as much as possible. After struggling with the traditional notions of validity, we concluded that, in order to legitimate the truth value of our research outcomes, the "Methods" make up the dimension that crosses both paradigms. Methods are those features that make up the content of a researcher's decisions, thus the label. The "Methods" box leads into the two paradigms, the "Qualitative" and "Quantitative" boxes.

We adopt the traditional issues for validity in quantitative research. These concerns about validity include both external and internal validity, on the one hand, and measurement validity, on the other hand. Both these categories of concern are generated by the need to have confidence that our test, data, or design does indeed measure or reflect or produce what we intend it to measure, reflect, or produce.

31

Our thinking about validity concerns in qualitative research positions us closest to what Denzin (1994) describes as the *postpositivist perspective*; that is, there needs to be a separate set of criteria for assessing qualitative research. The postpositivists call for a set of rules or procedures to establish validity. These criteria, from Lincoln and Guba (1985) originally (and as we conceptualize them), serve a similar purpose that criteria developed by Campbell and Stanley (1963) serve for quantitative research. In Figure 3 we present some of those criteria that methodologically support the legitimation of qualitative research. There are 14 methods, beginning with "Prolonged engagement." All are discussed in more detail later.

The concept of validity and the language used to discuss validity are not well established or agreed upon in the educational research literature. Some researchers seem to assume a dichotomy between the concern of qualitative researchers for validity and that of quantitative researchers for reliability (see Appendix A; Rist, 1977; see also J. K. Smith, 1985). Hammersley (1992) claims one type of truth (validity) is reliability. This argument is founded on different philosophies rather than on different methodologies. Our position is that validity cannot exist without reliability. By definition, validity estimates the extent to which the test or set of data or design actually measures or reflects or produces what it is supposed to measure, reflect, or produce. The basic purpose of reliability is to help researchers estimate validity as an estimate of measurement error (Newman & Newman, 1994). Therefore, if one has validity, there is no need for estimates of reliability.

Polkinghorne (1991) asks whether or not the findings of the study are believable. Then he defines validity as the correspondence between findings and "reality." Others, such as Lincoln and Guba (1985), prefer the term *credibility* to the term *validity*.

Our emphasis in exploring the validity of research methods across the qualitative-quantitative continuum is in enhancing the truth value of the outcomes of the research we conduct. In Figure 3 we show the reasoning that connects methods to the truth value of those outcomes.

Each "Methods" list leads to a point labeled "Design." All the methods together create the design. The methods are the design.

Enacting the design in each paradigm leads to the next point on the figure, "Research Outcomes," which are common to both paradigms. From the outcomes, finally, comes "Truth Value," our last point on the figure. The *truth value* label, while originating with qualitative research methodology, is synonymous with the validity of the research results. To some extent, it is what Polkinghorne is asking: are the findings of the study believable or true?

Validity Issues in Quantitative Research and Their Relationships to Methods

There is an important difference between measurement validity and design validity. The first, measurement validity, estimates how well the instrument measures what it purports to measure. The second, design validity, encompasses internal and external validity. Measurement validity falls under Campbell and Stanley's "instrumentation" (1963). Internal validity is the extent to which any causal difference in the dependent variable can be attributed to the independent variable. External validity is the extent to which the results of the research study can be generalized to other settings or groups.

Measurement and design validity are not independent. A research design is only internally valid if it has measurement validity and reliability. This also is true for the ability to generalize: estimates of the stability of the measurements are needed to estimate external validity. Measurement validity is a subset of internal validity. As discussed in the next section, strong measurement validity diminishes the threats to internal validity that come under Campbell and Stanley's category of "instrumentation." We chose to separate measurement validity and design validity to clarify the discussion.[1] Next we discuss internal and external validity as they relate to quantitative research; then we discuss validity in qualitative research.

Internal and External Validity

We need standards to evaluate research on at least three dimensions: the likelihood of the variables' being causally related; the confidence one would have in being able to generalize the results; and the level of consistency between the researcher's purpose, assumptions, methods, and conclusions. Design validity for quantitative research traditionally has been addressed through the concepts of internal and external validity. We, along with Goetz and LeCompte (1984), argue that a similar concern should be addressed for those studies that are predominantly on the qualitative end of the continuum. Within the context of our holistic view, however, we subsume all of these consistency questions into the concept of *methods* validity. As shown in Figure 3, the researcher makes decisions about methods, controlling for threats to internal and external validity, and the methods become the research design.

Internal validity is defined as "the extent that one can say the independent variable causes the effects of the dependent variables; in other words, one has the ability to assume causation to the extent that the researcher has control" (Newman & Newman, 1994). LeCompte and Goetz (1982) ask this question to get at internal validity: "Do scientific researchers actually observe or measure what they think they are observing or measuring?" (p. 43). To the extent that they are observing and measuring what they think they are, they have validity. The second conceptual area is external validity, defined as "the extent that a study is generalizable to other people, groups, investigations, etc." (Newman & Newman, 1994, p. 119). LeCompte and Goetz (1982) ask this question to assess external validity: "To what extent are the abstract constructs and postulates generated, refined, or tested by scientific researchers applicable across groups?" (p. 43).

Keppel (1973), for one, describes the great amount of effort among quantitative methodologists to achieve internal validity. Such researchers employ controls (i.e., holding constant as many factors as possible that may influence the phenomenon under study) to most efficiently measure the influence of the treatment. Tests are conducted under the same environmental conditions, using the same instruments, by the same researchers. Factors that are not controllable are allowed to vary randomly across treatment conditions, which works as another control mechanism. Randomization of what Keppel calls "nuisance variables" is a major way of obtaining internal validity, that is, "the elimination of biases which, if present, might invalidate any conclusions drawn concerning the manipulations of the experiment" (1973, p. 314). This randomization removes the effect of bias and variability due to such factors.

Other attempts to achieve internal validity include counterbalancing across the treatment conditions. Called *equivalent forms* (Newman & Newman, 1994), it involves taking two learning tasks, for instance, and giving Group 1 the tasks in *A–B* order and Group 2 the tasks in *B–A* order. In this case, the stimulus order is used as an independent variable, and its influence on the variability in the dependent variable can be isolated and measured. Measuring it assists the researcher in obtaining internal validity because its effects can be separated from the treatment effects.

Without internal validity, one can only conclude that the approach being used to answer the question of interest is capable of estimating the relationship, and no statement about causation is possible. (Later in this chapter we discuss qualitative methods.) Even though there are those among the ranks of qualitative researchers who say they are not interested in internal validity, those who wish to infer causal relationships must be concerned with this aspect of their research. In fact, even some who dismiss this concern as being only a quantitative researcher's dilemma will admit to processes like triangulation and theoretical sampling, which are conceptual attempts and techniques to get at internal validity.

External validity reflects the extent to which the design and the data match the world. While tight controls over the variables of interest increase internal validity, they tend to do so at the expense of external validity—that is, the more laboratory-like the conditions the more precise and valid are the measures. However, the world is not as tightly controlled, and the variables operate in the world outside the laboratory. Sampling of subjects across several strata that reflect the world to which the results will be generalized (e.g., age, socioeconomic status, occupation) increases external validity. The inability to control independent variables in ex post facto research (while it tends to decrease internal validity) tends to increase external validity because such variables are more relevant to their real-world distribution (Newman & Newman, 1994).

Diagrams used to illustrate research designs help the researcher plan, interpret, and analyze. Symbols, defined here, are used to depict the components of the research strategy and are applied in chapter 5. While the following are examples of symbols commonly used, they do not include all possibilities, nor are they universally used.

X	Treatment or experimental treatment; refers to the experimenter's treatment for a group
$-X$	No treatment or absence of treatment (one group might receive treatment, a second group not; second group is commonly called the *control group*)
(X)	Independent variable, not manipulated and either attribute or assigned variable
O	Measurement or observation, frequently some type of test score; can have any number of subscripts
R	Random assignment of subjects to groups
M	Assignment of subjects to groups using matching
M_r	Assignment of subjects to groups by first matching subjects and then randomly assigning each matched subject to group

Based on the classic work of Campbell and Stanley (1963), design validity is limited when an assortment of factors are uncontrolled.

History is a factor that could account for a change in a group. *History* means any nontreatment, extraneous event that intervenes between the pretest and posttest measurements. For example, a school district investigates the effect of a new set of mathematics textbooks on achievement by measuring achievement at the beginning and at the end of the year. During that year, however, all the students move to a new building. This is an example of a history factor that might affect the results of the study. The researcher, in this case, would not know if the new books or the new school caused any difference in student achievement outcomes measured in the study.

Another factor that could account for a change in a group is *maturation*, which means any growth or development that would normally take place independent of an experimental treatment. An example would be an analysis of the impact of a school lunch program on students' growth. Continued growth normally would occur as part of their maturation. In this case, the researcher's task is to identify what change is due to the lunch program and what change is due to ongoing maturation. History and maturation are often confused because each is related to something occurring in the time between the pretest and the posttest. The distinguishing point is that historical effects are external, while maturation effects are internal. Other examples of internal changes would be psychological, such as boredom, or physiological, such as fatigue.

Factors associated with measuring devices, *testing* factors, can also cause change to occur. This can happen when a pretest sensitizes people to an experimental treatment and actually causes them to behave differently during that treatment. Suppose a teacher is evaluating the impact of a character program on the moral development of students. If the teacher gives a pretest that asks questions about morality, the pretest could get the students concerned about the topic and thus influence them, making them more receptive to the program than they might have been without the pretest. This same phenomenon can take place with other factors, such as achievement. Testing effects are the results that the first test has in sensitizing subjects to the treatment, which then affects the posttest.

On the other hand, the term *instrumentation* refers to the effects that are due to unreliable measurement instruments. If one has an unreliable pretest measurement, any change noted in the posttest measurement might be due to the instability of the measurement device rather than to the treatment. Instrumentation is a threat to internal validity and is related to the validity and reliability of the data. The data the researcher uses as evidence are gathered through various instruments that measure the variables of interest. The next section describes issues of validity and reliability of such measurements.

Measurement Validity and Reliability: Instrumentation

Two concerns of the researcher when collecting data by means of a measuring instrument are the validity and reliability of the instrument. Weaknesses in either threaten internal and external validity in this instrumentation category. In this section, we discuss measurement validity first and follow with a discussion of reliability.

A test or measurement instrument has what is termed *face validity* to the extent that it appears to the individuals being assessed to be measuring what it purports to be measuring. This is generally considered to be a poor estimate of validity. It might be important because, if a test does not have face validity (credibility), it can decrease the likelihood of people participating or volunteering.

When experts in the content areas make subjective judgments about the validity of the instrument, there is said to be *expert-judge content validity*, which has also been called *logical validity* and, sometimes, *definitional validity*. It is similar to face validity, but it is generally estimated by using a table of specifications in which the purported content the test is supposed to measure is listed and assessed by how well and how completely the items represent content areas.

How well one assessment instrument correlates with an already-established, valid assessment instrument is known as *concurrent validity*. *Known-group validity* is a type of concurrent validity. It is estimated by how well the instrument differentiates between two known groups. If the instrument is supposed to measure successful marriage, the instrument should be able to distinguish between two groups of people who have been identified previously as successfully or not successfully married. To the extent the instrument can do this, it has known-group validity.

An estimate of how well an instrument predicts a future assessment outcome is called *predictive validity*. The major difference between predictive validity and concurrent validity is that concurrent validity tends to be occurring within the current time span, while predictive validity is future oriented. Sometimes concurrent and predictive validity are combined and are then called *statistical, empirical,* or *criterion validity*.

The most important and the most difficult to estimate is *construct validity*. It is used to estimate is how well the instrument is measuring the underlying construct it is attempting to measure, and it is generally estimated by the use of a combination of the other types of validity mentioned. A statistical technique called *factor analysis* also is used to estimate construct validity.

Quantitative research studies that lack documented measurement validity have limited truth value; and researchers need to remember that such validity, when it is claimed, is always an estimate, never an actual measure.

If validity is confirmed, having reliability is implicit; however, it is possible to have reliability without validity. The basic framework we are using is that the major purpose of reliability is either to support or to improve validity. Reliability describes consistency. While validity estimates how well a study or a set of instruments measures what it purports to measure, reliability estimates tell whether the outcomes will remain stable over time (i.e., whether they are "repeatable") or whether they are consistent among independent observers (i.e., whether different observers will report the same outcome).

In quantitative research, reliability in data collection is assured in three ways: measuring internal consistency, applying test-retest correlation coefficients, and using equivalent forms of the instrument. If reliability is not assured, then the scientific assumption of accuracy of measurement is violated. The facts are not repeatable.

Just as reliability is estimated by calculating the internal consistency of a test form, a similar measure can be derived from a *structured interview schedule*. Control over the timing, the environment, and the question order is possible where no such control is possible with questionnaires. To the extent that these controls enhance validity, they fulfill reliability requirements by definition. For nonstructured interviews, no such reliability estimates are possible.

One cannot have validity without reliability and, concomitantly, to the extent that one has validity one need not estimate reliability. For this reason, the predominant effort in this book is issues of validity.

Whenever groups are selected on the basis of extreme scores and for no other reason, a phenomenon called *statistical regression* occurs. If subjects are selected for a study solely because they score extremely low, at posttest they will tend to score higher, regardless of the treatment. The opposite is also true. Both are examples of extreme cases regressing toward the mean of the population. Thus, significant differences between pretest and posttest scores can occur because extremes were initially selected.

The loss of subjects between testings is called *experimental mortality*. If a sample of 100 subjects score an average of 95 on an IQ pretest and, for some reason, 50 subjects drop out before posttesting, then the average IQ of the remaining 50 might be 150. Therefore, the differences between pretest and posttest scores would, in all likelihood, not be due to the treatment but due to the differential loss from pretest to posttest (i.e., mortality).

When subjects are assigned to two or more comparison groups and not all groups are given the treatment, there is *selection bias*. If these groups are different before treatment, then any difference between pre- and posttest scores may be due to the initial differences rather than to the treatment. An example would be assigning individualized instruction (treatment) to highly motivated children and traditional instruction to children with little motivation. If the groups were tested for gains at the end of a unit, any difference found might, in fact, be due to initial motivation differences rather than to treatment differences.

These contaminating factors must be considered when evaluating research. They are useful criteria in judging the quality of research.

Ex Post Facto Research

There are two types of independent variables, *active* and *attribute*. Active variables are under the control of the researcher and can therefore be manipulated. Attribute variables, such as gender and race, cannot be manipulated. If all the independent variables are nonmanipulatable, then the research is defined as *ex post facto*. Ex post facto research is sometimes relegated to an inferior position among types of research design. The terms *ex post facto research* and *correlational research* are sometimes used interchangeably. Borg and Gall (1989) refer to this condition as "causal comparative" (p. 53). In correlational research, causation cannot be inferred. Many methodologists warn against possible misinterpretations of research in which the experimenter does not have control over the independent variables. Some consider ex post facto research to be exploratory (Newman & Newman, 1994).

In ex post facto research, causation is sometimes improperly inferred by those who assume that one variable is likely the cause of another because it precedes it or because one variable tends to be highly correlated with another (e.g., smoking—the independent variable as-

41

sumed to cause cancer—the dependent variable). While a correlated and preceding relationship is necessary, it is not sufficient for inferring a causal relationship. To assume a causal relationship, one must have internal validity; that is, all other explanations for the effect on the criterion (dependent variable) are controlled for, and the only possible explanation for changes in the dependent variable must be due to the independent variable under investigation.

Only with a true experimental design does one have the experimental control to achieve internal validity. Ex post facto research lacks this control for a variety of reasons. One cannot randomly assign and manipulate the independent variable because it has already occurred and is not under the control of the researcher. Another common weakness is that the design is not capable of controlling the confounding effects of self-selection. For example, suppose one conducts research to see what effect early childhood training has on motivation. Also suppose that a significant relationship is found between early independence training and later adult motivation. One might, therefore, incorrectly conclude that the independence training causes this adult motivation. Another explanation might be that volunteer subjects who have had independence training are more likely to demonstrate a greater degree of adult motivation. What causes this motivation might be more related to what causes the subjects to volunteer than to the independence training.

If the question deals with causation, clearly ex post facto research is inappropriate. However, if the question deals with relationships, it is appropriate. Sometimes a research question has independent variables that cannot be manipulated. In these cases the researcher can either decide to do ex post facto research or decide to do no research at all. One of the most effective uses of ex post facto research is in helping identify a small set of variables within a large set of variables related to the dependent variables for future experimental manipulation.

We have discussed threats to internal validity. Our discussion, derived from the classic by Campbell and Stanley (1963), now moves to external validity.

External validity, according to Campbell and Stanley (1963), addresses concerns for generalizability. Four threats to the generalizability of research findings are, firstly, the reactive effects of testing in which the pretest might cause the participants in the study to be either more or less sensitive to the treatment. Secondly, the reactive effects of the research situation itself threaten generalizability. In both of these cases, the sample of participants and their experiences become unlike the population to which the researcher wants to generalize. Thirdly, generalizability can be lessened when there is interaction between biases in the selection of participants and the treatment. Lastly, multiple-treatment interference can limit external validity as well. That research participants are subjected to more than one treatment can result in prior treatments having an effect on later treatments. The results of the later treatments cannot be generalized to groups who have not experienced the earlier treatments.

Our discussion of design validity in quantitative research has emphasized the two concerns of researchers when conducting experimental studies in the quantitative paradigm: the extent to which any difference detected in a study can be related to the independent variable (internal validity) and the extent to which results of the study are representative of the population, or are generalizable (external validity).

Before we move from these issues in the quantitative paradigm to similar issues in the qualitative paradigm, we briefly mention multivariate research. This segue into qualitative research is appropriate because, rather than investigating one variable at a time, multivariate research is based on the complexities of most human and social science research. More than one variable is almost always operating in the questions researchers ask. In the qualitative research situation, which we cover next, human complexity, at some level, underlies the research question that leads to qualitative methods.

Validity and Legitimation of Research

Univariate and Multivariate Analyses

Quantitative researchers use single dependent variables (univariate design) or multiple dependent variables (multivariate design). Attempting to represent the complexities of social phenomena leads to research questions with many variables. Factorial designs are one way to achieve external validity. Higher-order factorial designs (two-way, three-way, four-way, etc.) more closely approximate the external world than do one-factor designs.[2] In essence, each additional independent variable increases the design's relevance to reality as long as such variables account for some unique variance in the dependent variable. For example, the effects on math achievement are not explained merely by IQ but more credibly by IQ, family background, parents' education, and socioeconomic status. However, with each additional variable, the number of cases in each cell may decrease (the subject-to-variable ratio decreases). As the number of variables increases, one must consider the need for increasing the sample size. Therefore, some researchers use multivariate analysis as opposed to univariate analysis to increase external validity. In other words, behavioral phenomena almost always have an impact upon more than one dependent variable; and measuring one while accounting for the variance of others is more likely to result in conclusions that reflect relevance to the external world.

According to Michael Patton (1980, 1987, 1990), qualitative methodology is based on the assumption that the study of human behavior must be different from the study of nonhuman phenomena. Strike (1972, cited in Patton, 1980, p.44) who claims that, because humans have "purposes and emotions," they make plans, construct cultures and hold certain values and purposes. In short, a human being lives in a world that has meaning; and, because one's behavior has meaning, that meaning can be discovered and explained.

Patton appropriately infers that human behavior is much more complex than nonhuman behavior. While Patton argues for the qualitative paradigm, this view is not so different from a stance taken by quantitative researchers. Quantitative researchers indicate that, in attempting to try to understand even the simplest of human behavior, one must examine many variables and the situations in which they occur. In other words, variables may have differential effects, depending upon the specifics of the situation. They point to design strategies that control for multiple effects on dependent measures. Following Patton, one could test for the interactive effects of emotion and purpose on a certain behavior. Furthermore, "perceived ability," "likelihood of failure," and "need states" could be measured, and their relationships to behavior could be tested. While Patton implies that science can relate only to main-effects questions, one can, in fact, use advanced scientific design methods that consider the comprehensive milieu in which human behavior takes place.

Validity Issues in Qualitative Research and Their Relationships to Methods

Yvonna Lincoln (1995), in an article in the journal *Qualitative Inquiry*, presents the historical development of the dialogue about validity—the dilemma of criteria—in a naturalistic context. She places into this historical context the five criteria she and Egon Guba first described in 1985.

In our discussion of the philosophical underpinnings of qualitative and quantitative research at the beginning of this book, we noted that the notion of *truth* is problematic because truth becomes a social construct, idiosyncratic and situationally specific. Is validity a possible criterion given the nature of qualitative inquiry? In Lincoln's article, she outlines the responses to what Denzin (1994) has called the "crisis of legitimation" in qualitative research. Many qualitative methodologists and thinkers have responded to this question of validity (or truthfulness or legitimation), and a brief summary is presented here to put our response, the postpositivist perspective, in context. (We eventually return to Lincoln and adopt criteria she and Guba first articulated.)

Hammersley (1992) describes four philosophies or stances that call for different ways of thinking about assessing qualitative research. First, he describes "methodism," his synonym for positivism. Within this philosophy, the scientific method is a way to truth. Hammersley contends that in this view we cannot know anything that exists outside our experience. Validity cannot be connected to an external reality, so it is connected to what we usually think of as reliability "(agreement between the findings of different observers or between the findings of the same researcher on different occasions) and/or predictive validity (agreement between the results of the research and established measures of the relevant property)" (p. 196). He contrasts methodism and "realism." In this second philosophy, the researcher establishes validity because the researcher gets close to the subject under study. There is a strong correspondence between the knowledge gained in the research and the reality it represents. This philosophy assumes a reality out there that can be known. Third, because ethnographers construct the reality of the people or cultures they study, this research requires a validity that he calls "relativism." Relativism accepts multiple valid accounts of reality based on this constructivist epistemology. Consensus within a community about the truth of the findings constitutes validity in these cases. "Instrumentalism," his fourth philosophy, abandons validity concerns and opts, instead, for research that does some good. This notion suggests the philosophical stances of critical theorists and feminists.

As Hammersley moves within several notions of validity, the construct of truth and validity are questioned in a postmodern world. Multiple realities are acknowledged; and in a postmodern context, Tierney (1993) uses ethnographic fiction to explore organizational life. He cites Kitzinger to define the purposes of such a research strategy. "The attempt to induce in the reader a willing suspension of disbelief, while simultaneously acknowledging that one's argument is an 'account,' a 'construction,' or 'version,' rather than objective truth can take social constructionist writing into the realm of the arts" (Kitzinger, 1990,

p. 189). Tierney comments: "Through fiction, then, we rearrange facts, events, and identities in order to draw the reader into the story in a way that enables deeper understandings of individuals, organizations, or the events themselves" (p. 313).

Given this purpose and this method, validity assumes quite a different perspective. Tierney posits that validity then asks questions, such as: Are the characters believable? Is this situation plausible? Has the text led me to reflect on my own life?

Another qualitative researcher, Patti Lather (1993), poses "four post-modernist kinds of validity" as alternative ways of thinking about truth that reject correspondence theories of truth (positivism) (p. 54). "Ironic validity," "paralogical validity," "rhizomatic validity," and "embodied validity" question the notion of truth for postmodernists. Legitimation of qualitative research calls for ways to claim that the research has truth value, is trustworthy.

Lather's *ironic validity* fits the postmodernist epistemology because it sees truth as a problem. The truth value of the research lies in its ability to show us coexisting binaries, coexisting opposites (p. 57). *Paralogical validity* is that quality of research that legitimates because it reveals paradoxes, "undecidables," parts of meaning that are incapable of being categorized. Lather describes the legitimation of research that includes excerpts from interviews that are unmediated by the researcher, that cannot be interpreted, or that are not interpreted because of the researcher's unwillingness to diminish them (p. 57). *Rhizomatic validity*, unlike the deeply ingrained meanings that solidify into the postpositivists' categories through content analysis, fits the postmodern rejection of stable truth. This validity comes through "the crossings, overlaps, the meanings with no deep roots," the meaning that comes through mapping not merely through describing (p. 58). (We are reminded of path analysis in the quantitative paradigm: following where the data patterns take you in constructing a theory.) *Embodied validity*—validity that comes from the idiosyncratic nature of the study—comes from what Lather describes as the researcher knowing more than she is able to know, writing more than she is able to understand (p. 59). In essence, this is the nature of interpretation, bringing a sort of closure to the intellectual and emotional sorting and sifting of data.

Because the knowledge generated by qualitative research is the result of a social construction, Steinar Kvale (1995) claims this phenomenon is related to construct validity and is the goal of qualitative research. He presents three approaches to validity that are generated by this reasoning. Adopting the postmodernist perspective, the correspondence theory of truth does not apply. The social construction of reality is validated only through practice.

Kvale assumes that validity for the postmodernist must be different than for the positivist. Validity as a concept seems to imply a boundary line between truth and nontruth, according to Kvale. For the postmodernist, there can be no universal truth, but there may be "specific, local, personal, and community forms of truth, with a focus on daily life and local narrative" (1995, p. 21). More clearly than other writers, he explains the postmodern perspective.

The postmodern condition is characterized by a loss of belief in an objective world and an incredulity toward metanarratives of legitimation (Lyotard, 1984). With a delegitimation of global systems of thought there is no foundation to secure a universal and objective reality. The modern dichotomy of an objective reality distinct from subjective images is breaking down and is being replaced by a hyperreality of self-referential signs. There is a critique of the modernist search for foundational forms and its belief in a linear progress through more knowledge. The dichotomy of universal social laws and unique individual selves is replaced by the interaction of local networks, where the self becomes an ensemble of relations. The focus is on local context and on the social and linguistic construction of a perspectival reality where knowledge is validated through practice. (p. 24)

From Kvale's perspective, our qualitative-quantitative interactive continuum fits the postmodern philosophy. We assume no objective unitary truth about research methods and truth. We reject the dichotomy of universal laws of research. We do not, however, fit his three notions of validity within our thinking. Kvale refers to Polkinghorne's notions of validity: "Validation becomes the issue of choosing among competing and falsifiable interpretations, of examining and providing arguments for the relative credibility of alternative knowledge claims" (Polkinghorne, 1983, p. 26). Kvale labels his validity constructs *investigation validity, communicative validity*, and *action validity*.

Kvale's *investigation validity* is the quality of craftsmanship. It is the researcher's quality control. He lists checks that the researcher should make after interviewing, for example. The consistency of what a subject says in an interview is checked against other statements he or she makes. Others are interrogated, for example, in a kind of triangulation. Investigation validity includes how theories are derived from the data: how the researcher should begin to conceptualize the topic. For example, Kvale discusses research on student grades. Does the theory emerging from the data relate to assessing knowledge, or does it relate to the political dynamics of separating groups of students? To the extent that the theory aligns with the data and the purpose, one assumes validity.

Kvale's second validity is what he calls *communicative validity*. As he describes it,

> Communicative validity involves testing the validity of knowledge claims in a dialogue. Valid knowledge is not merely obtained by approximations to a given social reality; it involves a conversation about the social reality: What is a valid observation is decided through the argumentation of the participants in a discourse. (p. 30)

His third validity is *action validity*, in which the justification of the truth of the research is based on whether or not it works. Kvale refers to Patton (1990) and Patton's notion of credibility and likens it to his own action validity. According to Patton, the test of credibility of an evaluation report is whether or not it is used by decision makers. Truth, according to Kvale, is whatever is helpful in taking action to get a desired result. He makes comparisons to Polkinghorne (1991). Polkinghorne claims that the validity of case studies, narratives, and so on can be tested against their effects on practice. Kvale's final reference is to Lincoln and Guba (1985), who say that inquiry helps understanding, and through understanding, participants increase the control of their lives.

Lincoln and Guba, in their 1985 classic work, *Naturalistic Inquiry*, set forth criteria with which to assess the truth value of qualitative research. Goetz and LeCompte (1984) build on these ideas. Their works are foundational to much of our thinking and to the qualitative-quantitative interactive continuum.

Summary: Design Validity Criteria (Predominantly Qualitative)

What follows is a list of criteria to help one begin analyzing predominantly qualitative research for its design validity. Questions that might be used to probe the validity of methods in predominantly qualitative studies are included. These strategies are suggested as beginning points—not as an exhaustive list—and are taken from Guba and Lincoln (1982, 1989), as well as from Goetz and LeCompte (1984) and McMillan and James (1992).

1. Neutrality

How objective are the data? Given that no data collection is entirely objective, the reader must see where this research falls on the objective-subjective continuum. The reader must look at the biased or unbiased nature of the inquiries. Are the data available for the public to see? Are judgments documented with evidence, as opposed to being merely the author's opinions? Is there more than one observer? If so, is there consistency between observers? Is there consistency in the interpretation of the data?

2. Prolonged Engagement On-site

Did the author observe long enough to get an accurate reflection of the culture or history? If only one observation is taken, it will reflect one small portion and will not capture the essence of the culture or the situation. An example of this would be a visitor from outer space coming to Ohio during the winter and reporting that there were trees without leaves. Another outer-space visitor arriving in summer would report trees with leaves. Both would be accurate in their observations, but neither would accurately reflect the actual situation.

3. Persistent (Consistent) Observation

Was sufficient time spent on-site to get an adequate picture of consistency of behavior? Was the observation typical or something that does not usually occur? The research report should reflect this. Even though this sounds similar to prolonged engagement, it is different in that the purpose of prolonged engagement is to be able to detect cultural trends or idiosyncrasies, while the purpose of persistent observation is to identify or estimate if a particular behavior is (or sets of behaviors are) frequent or infrequent.

4. Peer Debriefing

Did the researcher talk with any other professional to get another perspective on what he or she saw or experienced? Debriefing is somewhat like controlling for countertransference in psychoanalytic terms. That is, one may get too attached to the environment in which one is trying to be an objective observer. Researchers may begin to interpret things from their own need base. They need other professional and expert interpretations to give feedback.

5. Triangulation

Did the researcher attempt to obtain a variety of data sources (e.g., different observers or different written histories)? If so, was there shared reality? While an important concept in much of the qualitative methodological literature, it might be considered somewhat quantitative in that one is looking for consistencies in perceptions. To some extent, triangulation might be looked at as a reliability check—but not always. It is possible that one source of data could be much more important than other sources in understanding a particular phenomenon. Generally, however, the more sources one looks at the more likely one is to have a complete perception of the phenomenon.

6. Member Checking

Member checking refers to how accurate the data are. Were the data and interpretations continuously checked? One way of estimating the accuracy of personal observations is to check out those observations on members of the group one is observing. That is, when a researcher returns to those people interviewed and checks to make sure he or she "got it right," the researcher is member checking.

7. Referential Materials

Did the researcher use enough supportive material (e.g., documented recordings, readings, archives, or other materials that are available to others)? It is important for the researcher to document references, records, and interviews used and to let evaluators know how aware the investigator is of these materials. It is also important to indicate which sources were used in what ways and, if any available sources were not used, why they were not.

8. Structural Relationships

Is there logical consistency between different data sets? When attempting to interpret data and formalize conclusions, the researcher should support these insights, to the extent possible, by interweaving different data sets, which may come from different perspectives while supporting the common underlying and emerging meaning.

9. Theoretical Sampling

Did the researcher follow where the data led? The researcher typically enters the field and begins immediately to collect data. While data are being gathered, the investigator has begun to form explanations of their meaning. These tentative explanations (theory) suggest other data sources. In other words, the sampling of data in qualitative research is determined by the existing data (Goetz & LeCompte, 1984). The researcher attempts to capture the best theory that explains the data. A quantitative researcher might call this *soft hypothesis testing*. The researcher may change direction or collect different and/or additional data. On the other hand, such data sampling may provide supportive and corroborative interpretations of initial emerging theory.

10. Leaving an Audit Trail

Does the researcher have good documentation, so that another researcher can easily replicate the research? This not only means that someone would be able to replicate the current study but be able to confirm or to contradict the interpretation based on the same data.

11. Generalizability

That one should be able to generalize underlies science. However, we are unwilling to accept fully that generalizability is consistent with the qualitative paradigm. We have claimed throughout this book that, in principle, generalizability is the purpose of quantitative—not qualitative—research. In fact, we have assumed that, if the purpose of the research is to generalize, one should employ quantitative methodology. There seems to be growing interest among qualitative researchers in being able to generalize, even though it seems to be in violation of basic assumptions of naturalistic philosophy. Polkinghorne (1991), a qualitative methodologist, distinguishes between two types of generalizability: *statistical* and *aggregate*. The statistical model is more consistent with quantitative assumptions; and the aggregate model, based upon deep descriptors, is more consistent with qualitative assumptions. The deep descriptors are sufficiently comprehensive to allow the qualitative researcher to generalize to each and every member of the population. Donmoyer (1990) too states that *generalizability* can be broader than traditionally defined. He accepts traditional generalizability for statistical research (quantitative) and develops schema theory based on Piaget's notion of assimilation, accomodation, integration, and differentiation (qualitative research) (p. 197). The growing body of published qualitative research has put pressure on research methodologists to create ways in which results of such studies can be applied to wider audiences, or generalized. The following concepts, *applicability, transferability (context limited),* and *replicability,* are examples of the kinds of questions to ask to improve those efforts.

APPLICABILITY

Can this research be applied to other samples? The reader must look at the sample size as well as the sample characteristics. The important assumptions are that the hypothesis is an emerging one and not a sample-to-population statement, that there is no test of significance, and that the purpose of deep descriptors is to describe in detail the characteristics of the sample being investigated so that others can make logical judgments about whether the sample is comparable to other samples. To the extent that the samples are similar, applying the results can be done comfortably.

CONTEXT LIMITED (TRANSFERABILITY)

Do the findings of the research hold up in other settings or situations? To the extent that it can be argued from a logical or data-based point of view that what is being observed or accomplished is not dependent on the context in which it was observed, that it is not context limited, it can be said to be transferable to other contexts, and it is therefore generalizable. For example, the effects of praise on a student may be more context independent than context specific; that is, the effects of praise may affect academics, as well as performance in sports and in social situations.

REPLICABILITY (CONSISTENCY)

What is the likelihood that a given outcome or event will happen again if given the same circumstances? Replicability is difficult to accomplish with any level of confidence, especially in a natural setting. One must identify changes that are due to identified effects and the frequency of these common occurrences at different points in time, in different settings, by different observers. When these data are available, they are valuable and provide important insights.

12. Negative Case Analysis

Has the researcher taken into account all known cases? Continually revising the emerging hypothesis until all known data are explained by the hypothesis is the concept of negative case analysis. In essence, this is a sequence of expanding and reshaping one's interpretation until all outliers are included. A quantitative researcher's predisposition is to avoid including what appear to be chance variations in the data.

13. Truth Value (Credibility)

What confidence does the reader have in the findings of the research? We previously described the concept of construct validity, claiming that an instrument has construct validity to the extent that it has all other types of validity (i.e., face, content, expert judge, concurrent, and predictive). Similarly, a study has truth value to the extent that the above 12 components exist. Any one study is unlikely to have all of the components; and, for all studies, some of these components are more important than others; but, generally, the more components the greater the truth value.

4

Strategies to Enhance
Validity and Legitimation

RESEARCH DESIGN IN QUANTITATIVE RESEARCH is made up of the methods one selects to carry out the study. Similar situations exist in qualitative research. The methods become the design—our focus in chapter 3. We continue the discussion of legitimacy of design by focusing in this chapter on the qualitative paradigm. While ways to mitigate threats to the validity of quantitative research are well recognized, ways to mitigate threats to qualitative research are not universally accepted. From our postpositivist perspective, we present ways to enhance the design validity of studies that are predominantly conducted to answer questions calling for the qualitative paradigm.

At the conclusion of this chapter, the reader should be able to

1. Describe ways to ensure the validity of observational methods
2. Describe ways to ensure the validity of grounded theory methods
3. Describe ways to ensure the validity of case-study methods
4. Describe ways to ensure the validity of interviewing methods
5. Describe ways to ensure the validity of historical methods
6. Describe ways to ensure the validity of ethnographic research
7. Describe ways to ensure the validity of phenomenological research
8. Describe triangulation and its effects on truth value

Observational Methods

Observation is the most frequent data-collection method used in qualitative research (Becker & Geer, 1960; Lofland, 1971; McMillan & James, 1992). K. D. Bailey (1978) asserts that "all other things being equal," observation has greater face validity than "a second-hand account gathered either through interviewing or document study" (p.242). Face validity however, is a low-level estimate of validity, appropriate only as a last resort or when no other validity estimates can be obtained, for three reasons. First, *participant observation* (in which the observer is obvious to and involved with the subjects), claims Bailey, is less valid than a questionnaire would be for sensitive data. Second, the observer's expectations affect what he or she sees and reports, reducing the validity of the data. Third, and even more complex, is the lack of expectations that results when no structure is a priori given to the observer. Validity is thus diminished when the observer reports seeing either "everything" or "nothing." Bailey remarks, "It is clear that one can easily see what one expects to see even if it is not there, thus causing bias; this is an example of selective perception. However, the opposite—the complete lack of expectation of what is to be observed—can also lead to invalidity" (p. 243).

One other caution exists: the observer's sense perceptions are not always accurate. Particularly in a nonstructured observation situation, the element of surprise can dominate the sensory input of the observer, rendering reported data invalid.

Gay (1987) and Mouly (1970) discuss the potential invalidity of observational data when they call for, at the very least, a scientific basis for the observation. Mouly, more specifically, agrees that both the scientist and the layman observe, "but the scientist starts with a hypothesis and arranges the conditions of his observations to avoid distortions" (p. 282). He warns further about invalidity, especially of participant-observation techniques.

As the participant observer adapts more and more to his role as a participating member of the group, he becomes increasingly blinded to the peculiarities he is supposed to observe. As a result, he is less likely to note what would be significant to a more objective observer. As he develops friendships with the members of the group, he is also likely to lose his objectivity, and, along with it, his accuracy in rating things as they are. (p. 289)

Despite these pitfalls, there is validity in using the observational method for study of some phenomena, such as nonverbal behaviors.

All validity concerns described here affect both participant and nonparticipant observations. In participant observation, the researcher is a regular participant in the activities being observed; while in nonparticipant observations, the researcher is not a participant in the ongoing activities being observed. Compared to *participant*-observation strategies, the validity of *"nonparticipant*-observation strategies is greater because there is no reactivity among the subjects to the presence of the researcher. This reduction in bias, however, does not cancel out the other biasing (invalidating) effects.

Despite these limits on the validity of observational methods, some maintain that it is, nevertheless, a highly appropriate technique (e.g., Hakim, 1987). Lofland (1971), for one, designates a first priority to the observer's understanding of the subject's point of view when he asserts, "In order to capture participants 'in their own terms,' one must learn their categories for rendering explicable and coherent the flux of raw reality. That, indeed, is the first principle of qualitative analysis" (p. 7).

While this "understanding of the subject's point of view" is highly regarded, the statements may well be describing only observer bias. However, Becker and Geer (1960), as well as Lombard (1991) and Lincoln and Guba (1985), place the methodology in even higher esteem when they state that participant observation is the "most comprehensive of all types of research strategies." In detail, Becker and Geer (1960) maintain:

> The most complete form of the sociological datum, after all, is the form in which the participant observer gathers it: an observation of some social event, the events which precede and follow it, and explanations of its meaning by participants and spectators, before, during, and after its occurrence. Such a datum gives us more information about the event under study than data gathered by any other sociological method. (p. 133)

The observer's attention to a setting is described as an evolving role by Boostrom (1994). From his own experience, he shows how the qualitative researcher can move from an "almost inert receiver of visual and aural stimuli" to being interactive in constructing the account of what he or she sees. He sees this role change as a move through the roles of "videocamera, playgoer, evaluator, subjective inquirer, insider, and, finally, reflective interpreter" (p. 53).

Enhancing Validity of Observational Methods

First, because the subjective bias of the observer affects his or her reporting, having several observers from several backgrounds (or points of view) report on the same phenomena can increase validity. Coalescing their data reduces sensory-deficiency and misinterpretation error. Second, structuring the observation increases validity by focusing the attention of the observers on certain characteristics and events. Third, placing the observation on a scientific foundation by stating a hypothesis up front increases validity by avoiding distortion. Fourth, nonparticipant observation, as opposed to participant observation, increases validity. And, fifth, using observation only for studying those phenomena that are appropriate to this method (e.g., nonverbal behaviors and social interactions) increases validity.

60

Grounded Theory Methods

Observation is the first and key data-collection strategy of the grounded theorists. We cannot fully shift from discussing the observation methods in this section. To relate validity concerns to the grounded theory methodology, we first need to review the approach, described in the classic work by Glaser and Strauss (1967). The observers enter the research situation with no hypothesis. They describe what goes on and from the observational data they develop a hypothesis. From her study of the chronically ill, Charmaz (1983) elaborates:

Grounded theorists shape their data collection from their analytic interpretations and discoveries, and, therefore, sharpen their observations. Additionally, they check and fill out emerging ideas by collecting further data. These strategies serve to strengthen both the quality of the data and the ideas developed from it. (p. 110)

Second, grounded theorists simultaneously address the process of research and the product of research; they are inseparable. As information emerges from the data, it is put into an original theoretical framework.

Third, grounded theorists do not use traditional quantitative methods to verify the data. They "check their developing ideas with further specific observation" (Charmaz, 1983, p. 110) and may do additional observations in other settings or on other issues.

Data are analyzed by initial and focused coding techniques. *Initial codes* are constructed after studying the data. These codes are derived from the participants and the roles they play, the context, the timing and structuring of events, and the issues that are the focus of the participants' behaviors and interactions. Connections among all these phenomena are attended to as the researcher codes the data.

The next phase consists of *focused coding*. Here the initial codes (labels) are combined and raised to an analytical level and then put into categories. Sometimes categories are developed in the language of the clients and sometimes the researcher devises categories in his or her own terms. Categories can be broken up and recombined. The researcher may go to the literature to expand and clarify the codes and categories. In these instances, researchers use the literature as "a source of questions and comparisons rather than as a measure of truth" (Charmaz, 1983, p. 117). Grounded theory emphasizes process. In fact, categories, once developed, are not treated singly but are woven together to make meaning. This is a processual analysis.

Enhancing the Validity of Grounded Theory Methods

Data collected and results formed in grounded theory can be made more valid by use of the aforementioned alterations to observational techniques. It is immediately apparent, however, that beginning with a hypothesis at the outset of data collection violates the first and most important assumption of this method. Yet the initial-coding stage and the focused-coding stage of the process are not unlike an empirical researcher's coding of open-ended questions. This similarity reveals an area of overlap in qualitative and quantitative methods. A common process for the empirical researcher is to collect responses on a questionnaire, review all responses to a particular inquiry, and then categorize them. The resulting categories are then used as categorical variables in a statistical analysis. This traditional approach does not include, however, the ambiguous process of sifting and resifting, considering what information was left out, as well as what was said, and other subjective manipulations of responses. In fact, the processes described in the Charmaz chapter imply a highly skilled, insightful coder who has an unusually profound understanding of the psychology of human motivation, incentive, and behavior.

The theory-building purpose of the grounded theory method rests on underlying assumptions about procedures that are highly questionable, considering the definitions of validity assumed here. Therefore, initial coding may be misleading, and the resulting theory based upon the coding may be unreliable.

This description of grounded theory methods provides a fairly accurate picture of what the method involves, but it includes nothing about validity concerns. Therefore, several comments seem appropriate, as validity is the issue in this section. To begin, no hypothesis directs the data collection. As with nonstructured observations and unstructured interviews, validity might be diminished when the researcher is potentially bound only by his or her biases. Because the grounded theorists maintain an "independent" view of the data, untainted by even a literature review, it seems difficult to accept that their perceptions are bias free. This lack of the researchers' acknowledging factual assumptions at the outset, which is applauded as freeing the researchers to experience solely the constructs emerging from the data, can be seen as merely allowing the selective and subjective perceptions of the researchers full rein. That they go so far as to acknowledge that the data they collect may not even be related to the topic under study is a clear indication of the design's potential for invalidity.

Collecting "data" in a grounded theory study requires highly sophisticated researchers. To be able to attend to the context, the participants and their roles, the timing and structure of observed events, the connections between individuals and their problems, and the individuals' interpretations of their own situations while at the same time coding what the subjects fail to say, what they lack, what they gloss over, what they ignore, as well as interpreting their imagery making and their feelings, indeed, all require superhuman coding skills.

Collected data are assumed to be gathered by individuals with highly developed and comprehensive observation skills, and, to the extent that such assumptions are not met, the resultant data are invalid. Furthermore, that such observers are able to discern what the events *mean* to the participants, as Charmaz asserts, implies some sophisticated skills in interpretation as well.

Both simultaneous with and subsequent to focused coding is the process called *memo writing*. As described in detail by Charmaz (1983),

> Memos are written elaborations of ideas about the data and the coded categories. Memos represent the development of codes from which they are derived. An intermediate step between coding and writing the first draft of the analysis, memo writing then connects the bare bones analytic framework that coding provides with the polished ideas in the finished draft. By making memos systematically while coding, the researcher fills out and builds the categories. Thus, the researcher constructs the form and substance toward a finished piece of work and develops the depth and scope of the materials. . . . Sorting and integrating memos follows memo writing. These two steps may themselves spark new ideas which, in turn, lead to more memos. (pp. 120–121)

Sorting leads to memo writing. The integration of the memos follows from their sorting because, as must be repeated, the purpose of grounded theory is to develop theory. Integrating the memos helps develop the relationship (i.e., helps develop the theory). A problem, however, can evolve. Because the researcher starts with nothing—no theory, no hypothesis—he or she has no limitations or operational definitions of variables to determine what data should be collected. So one may find as the theory "emerges" that they need more data or need to define additional variables. This additional data collection is called *theoretical sampling* because it is "sampling [of data] aimed toward the development of the emerging theory" (Charmaz, 1983, p. 124). Its development comes from the inductive process and is evoked by coding and memo writing. The need for it means that the codes (conceptually) and relationships described (memos) have become sufficiently developed that the researcher can examine them in more depth. Theoretical sampling is a way to check those categories and relationships.

The weaving and sifting of categories of variables to formulate the relationships among them allows for, at least, a claim of subjectivity on the part of the researcher and, at most, a gross misinterpretation of actual facts. The grounded theorists accuse the empiricists of imposing a priori rating-scale values and codes on the subjects' responses, while it may be that the grounded theorists' own processual analysis is even more firmly based on researcher bias.

Case-Study Methods

The case-study method is one more design strategy under the qualitative rubric. Case studies can be single-subject designs or based on a single program, unit, or school. Merriam (1988) describes how to do case-study research, beginning with translating the research question into more specific and researchable problems, followed by techniques and examples of how to collect, organize, and report case-study data. In addition, she argues that case study is a helpful procedure when one is interested in such things as diagnosing learning problems, undertaking teaching evaluations, or evaluating policy.

Consistent with assumptions of qualitative research philosophy, the critical emphasis in case studies is revealing the meaning of phenomena for the participants. Stake (1981) acknowledges this assumption, claiming that case-study knowledge is concrete, contextual, and interpreted through the reader's experience. He prefers case-study methods because of their epistemological similarity to a reader's experience. He particularly notes the reasonableness of assuming the natural appeal of the case approach.

Case-study data come from strategies of information collection that have been described in Figure 2: interviews, observations, documents, and historical records. Patton (1980, p. 304) describes the three steps in conducting a case study:

1. Assemble raw case data

2. Construct case record

3. Write case-study narrative

Stake, a well-known advocate of naturalistic inquiry, considers validity to be an advantage of case studies because of their compatibility with reader understanding; in other words, they seem natural (1981). The validity limitations that have already been put forth in this book related to observational data apply to case studies as well. However, the counterbalancing of information from documents with data from observation and interviews strengthens the resulting validity. Invalidity of one set of data can be checked by conflicting or supporting results from the other sources, which is a type of triangulation (a concept discussed in more detail later in this chapter).

Enhancing Validity of Case-Study Methods

Case-study methodology has potential for increased validity for several reasons. First, because multiple data-collection techniques are used (e.g., interview, document study, observation, and quantitative statistical analysis), the weaknesses of each can be counterbalanced by the strengths of the others. Conclusions related to a certain aspect of a phenomenon under study need not be based solely on one data source. Second, validity may be increased by checking the interpretation of information with experts. Third, with case studies there are generally a variety of data sources. There should be a structural relationship among these sources. To the extent that these findings are consistent within the case, the validity is enhanced. Conceptually, this is similar to giving a battery of tests to obtain an estimate of consistency in the underlying constructs. Fourth, using a scientific method in which one hypothesizes something about the case and collects data to determine if the hypothesis should be rejected could add to validity and also help future researchers determine starting places for their research. All of these approaches would tend to improve understanding of the case and give in-depth descriptive information.

Interviewing Methods

Patton (1990) characterizes the research interview as a strategy to find out from people things that we cannot directly observe. Interviews can be structured (standardized) and unstructured (nonstandardized). Newman (1976) includes a third type, the partially structured interview. The structured interview is designed to collect the same data from each respondent, while the unstructured interview may be used to identify broader issues. In this latter case, each respondent may contribute a different perspective, depending on his or her position regarding the phenomena under study. Unstructured interviews "are totally dependent on the skill and training of the interviewer" (Newman, 1976, p. 13). To the extent that such skills are evident, the data collected are likely to be valid.

Structured interviews and partially structured interviews can be subjected to validity checks similar to those used in evaluating questionnaires. That is, are the questions consistent with the purpose of the study? The interview schedule (list of questions) or interview guide is created to direct the interview on a path consistent with the purpose. Diversity of opinion exists about the leeway a researcher may use with the interview guide.

Patton (1980) feels that the guide merely provides

topics or subject areas within which the interviewer is free to explore, probe, and ask questions that will elucidate and illuminate that particular subject. Thus, the interviewer remains free to build a conversation within a particular subject area, to work questions spontaneously, and to establish a conversational style—but with the focus on a particular subject that has been predetermined. Interview guides can be developed in more or less detail depending on the extent to which the researcher is able to specify important issues in advance and the extent to which it is felt that a particular sequence of questions is important to ask . . . deciding how best to use the limited time available in an interview situation. (pp. 200–201)

Others restrict the guide to a list of questions with a less free-wheeling attitude. Hakim (1987) maintains that the validity of this strategy is quite good. As research strategies, interviews provide both more complete and more accurate information than other techniques. Schatzman and Strauss (1973) consider it valid because they assume that all conversation between the researcher and others at the site is a form of interviewing. This "naturalness" lends validity to the information obtained. Spradley (1979) might contend that validity cannot be assumed but rests on the quality of the interviewing process the researcher employs. His seminal work, *The Ethnographic Interview,* is the most often cited source for planning interview strategies.[1]

Through probes, follow-up questions, and attention to nonverbal cues, the researcher is able to enhance the data collected. The data are valid to the extent the researcher is able effectively to execute these tasks. Limitations to validity exist, as with other qualitative methods, when the subjective bias of the interviewer affects the interpretation of the data in ways that misrepresent the subjects' reality. These invalidations may be more likely with the unstructured interview than with the structured one.

In fact, Mouly (1970) points to interviewer bias as the "major weakness" of the method. And so, while some claim that the ability to depart from a rigid structure, probe, and follow-up *increases* validity, Mouly would differ.

68

To the extent that the interviewer is allowed to vary his approach to fit the occasion, he is likely not only to complicate the interpretation of his results but, even more serious, to project his own personality into the situation, and, thus, influence the responses he received. (pp. 266–267)

Enhancing Validity of Interview Methods

Interviewing can be made a more valid technique in several ways. First, when structured, the questions can be checked against the objectives of the study. Second, a high level of interviewer training increases validity. Third, having several interviewers randomly assigned to subjects reduces error by spreading bias throughout the sample. Fourth, checking for consistency across subjects increases reliability, which adds to validity. Fifth, debriefing the interviewers after data collection also can help increase validity. In this process the researcher is able to check for interviewer bias and consider its effects on the interviewer's findings.

One definition of validity has to do with its ability to predict and explain underlying constructs. Once interview data have been collected, one can determine how well the interview explains certain underlying constructs related to the purpose of the interview. If one has hypotheses or assumptions to start with, the data can be used to see if these assumptions are verified (predicted) or contradicted. Based on these new findings, either the theory is supported and new assumptions formed, or new directions for future research are suggested, or both.

As is apparent from this discussion, the validity of qualitative methods frequently is increased by using what are considered to be traditional quantitative methods. From our perspective, this fact adds strength to the argument that the qualitative-quantitative dichotomy is false and supports the holistic concept of an interactive continuum.

Historical Methods

Historical research is carried out primarily through document study, as well as with interview techniques. There is disagreement about the scientific status of historical research methods, and several of these points of view are reviewed in this section. Some maintain that at all times the historian operates inductively, drawing from data to formulate conclusions as the research is carried out. No hypothesis is initially stated, according to Mouly (1970). Mouly further adds that historians frequently have to reconstruct the facts from unverifiable sources. These facts are based on their plausibility and can only be inferred; they cannot be measured. Furthermore, because the design necessarily focuses on unique events, one is unable to generalize. Shafer (1974), in an interesting treatise detailed later in this section, maintains that there are facts that we do accept from a historical perspective, and he discusses the historian's search for causation and the place of statistical analysis.

Good (1963) asserts,

> The historian thinks of the method of investigation as scientific, and the manner of presentation as belonging to the realm of art. . . . These interrelationships of history and science make it important that the modern historian be well grounded in the natural sciences. (p. 181)

One would expect that historians of Good's description would consider and deal with threats to validity. He assumes that historiography is a science, a quite different assumption than Mouly's. Good says, "History qualifies as science in the sense that its methods of inquiry are critical and objective, and that the results are accepted as organized knowledge by a consensus of trained investigators" (p. 181).

In fact, as Good describes the process of historical research, it does not differ substantially from the processes of quantitative research. However, he says, historians and quantitative researchers start at different places. Historians do not direct observations or experimentation, but they do utilize "reports of observation that cannot be repeated." As Good describes that process,

> historians cannot recall the actors of the past to reproduce the famous scenes of history on the stage of today. . . . Therefore, the historical method involves a procedure supplementary to observation, a process by which the historian seeks to test the truthfulness of the reports of observations made by others. (p. 183)

Even more closely aligning the historians with quantitative researchers and with the traditional scientific method, Good compares the two and then describes the uniqueness of the historian.

> Both historian and scientist examine data, formulate hypotheses, and test the hypotheses against the evidence until acceptable conclusions are reached. A number of historians, in emphasizing the interpretation and meaning of facts, have sought to identify tendencies, themes, patterns, and laws of history, while some of these investigators have dealt with such philosophical or theoretical problems in history as discovery of laws, unity and continuity, possibility or impossibility of prediction, and oversimplification growing out of the search for clues or keys. (p. 183)

In his discussion of historical strategies, Shafer (1974) acknowledges the bridge over the qualitative-quantitative gap made by the historian's use of data-analysis methods. He uses as his example the 1964 black vote in his description.

Dealing with past events that never exactly repeat themselves, he [the researcher] cannot conduct controlled experiments. Yet under certain circumstances he can apply statistical tools. Suppose, for example, his problem is whether the Negro vote was the principal cause of Johnson's victory over Goldwater in the State of Virginia in the 1964 presidential election. The researcher could probably design a study based upon comparison of voting districts with high Negro registration with districts of low Negro registration, working in a comparison between this election and earlier ones as well. The result of his investigation might well be a convincing set of correlations between the percentage of Negro registrants in a district and the amount of Johnson's majority in that district. Statistical analysis would then tend to support the hypothesis of the decisiveness of the Negro vote. (p. 51)

Thus, what would traditionally be considered a quantitative technique appropriately fits this research and adds validity to the study, demonstrating another advantage to denying the methodological purity of either end of the qualitative-quantitative continuum.

Mouly's assertion of nonverifiability of historical facts is tempered somewhat if the investigator has access to primary sources (i.e., original and factual documents). Invalidity originates from the secondhand nature of data; and, to the extent that original sources are available, the invalidity is reduced. Also, checking facts when possible with multiple sources would reduce the invalidity of this method.

Many (e.g., Mouly, 1970; Shafer, 1974; Wiersma, 1980) describe the processes of external criticism and internal criticism as steps in the historical researcher's work. Briefly stated, while these processes overlap, *external criticism* (called a *validity check* by Wiersma) deals with the genuineness or authenticity of the document being used. This is a step prior to *internal criticism* (i.e., the determination of the meaning and trustworthiness of statements within the document). The distinction is necessary because, while an original source is authentic, statements within it may not be completely accurate; they may be over-laid with the author's bias or political stance. Because document study is a predominant aspect of historical methods, these steps are critical. Shafer adds a third step, *synthesis* (i.e., the blending of evidence resulting from the first two steps to report historical events with accuracy). It is worth noting that Shafer admits that, while external criticism and internal criticism are validity processes, this third step is "necessarily riddled with subjectivism" (1974, p. 26). One must hope that it is not riddled enough to invalidate the outcomes of steps one and two.

Shafer (1974) acknowledges two stances regarding the historian's place on the qualitative-quantitative continuum: Mouly's perspective (history as a nonscientific mode) and Good's perspective (history as a scientific mode). He describes these two points of view as follows:

> One is that the historian's cultural experience or environment affects his interpretation of evidence on human affairs, and that as a result interpretations of history necessarily vary with the social environments of historians. The other is that inference (i.e., supplying data not explicitly provided by the contemporary writers or artifacts) must be indulged in with great care. (pp. 12–13)

In other words, the validity of historical methods is based on two assumptions: that the interpretation of history varies with the subjective social experience of the historian and that the reporting of history should not go beyond the database. Some moderation of this relativistic dichotomy of historical research occurred in the mid-1900s, according to Shafer, when, in essence, it was recognized that there are factual aspects of societal life that we can know and know with confidence. Therefore, the work of historians need not always be questioned (by the scientifically based) on validity grounds. Indeed, this epistemological conclusion assigned a new level of validity to historical findings that quantitative researchers could accept. As Shafer (1974) explains it,

On the theoretical level, relativism in the middle years of
the twentieth century was rather modified by arguments by
epistemologists and others that human activity shows some
"probabilistic" regularities, permitting assumptions and
explanations in which we may repose considerable
confidence. (p. 13)

Some purist quantitative researchers might read that conclusion
and see it as the historian's affirmation that human behavior, as well
as social phenomena, is lawful, measurable, and based on a scientifically discoverable theory that can be revealed by controlled designs
for hypothesis testing. And this is exactly what the quantitative researchers have been trying to convince the qualitative researchers of
all along. In any event, this conclusion by some—that historical research now subsumes the quantitative philosophy—should not be used
as a rationale to be negligent in estimating validity when utilizing historical methods.

Shafer goes to considerable lengths to detail the threats to the validity of historical research and the nuances of analysis and reporting
to which one must attend. Three cautions in particular are relevant
here. First, while historians struggle to delineate causes of events in
the past, causation can never be concluded. Historians must examine immediate precipitating events, as well as underlying events, that
may be related to the historical phenomena under study. Second, correlations may be found between precipitating events and the event
being studied. Third, he acknowledges that multiple causation is more
relevant to history than unidimensional causation. Not surprisingly,
these three tenets, explained at great length by Shafer (1974, p.
171–200), also are three of the basic tenets of quantitative research:
(1) causation can rarely be proven; (2) *correlational* is not synonymous with *causal*; and (3) multivariate relationships between cause-
and-effect variables are more common than univariate relationships.

In summary, historical research not only encompasses some of the methods already described (interview, case study) but also utilizes document study as a major strategy. Historians must seriously consider validity and its threats, as do all researchers. Utilization of data-analysis techniques in a traditionally quantitative sense is often useful in increasing validity.

Enhancing Validity of Historical Methods

One of the major problems with historical research is its inherent validity weakness, inherent because events have occurred in the past outside of the possible experiences of the researcher or the subjects, and so verification is always of a secondhand nature. Historical study can be made more valid by including checks on sources of data and ensuring that multiple sources are used.

Ödman, a professor at the University of Stockholm, studied witch trials in Stockholm, Sweden, during the 1670s. The reader is directed to his example of applying his model of interpreting historical evidence in an issue of *Qualitative Studies in Education.* He discusses the validity problem in historical interpretation, grounding his ideas in the work of Trankell (1972) and Ricoeur (1988). Ödman's (1992) discussion delves into the situations in which actual physical reality exists, as well as situations in which the researcher works at a symbolic level. Ödman's model should be reviewed by those pursuing historical research in education. The reference here to his work is insufficient for understanding it in depth.

To the issue of validity, Ödman calls on his predecessors for criteria:

How the validity problem is approached in an individual case depends partly on which kind of reality the interpretation is concerned with. . . . Two criteria which Arne Trankell formulates in his *Reliability of Evidence* (1972) are useful. . . . The first one can be summarized: *If an interpretation leaves an essential part of the information unexplained, this inter-pretation cannot be accepted as a valid description of the reality which the data are referring to.* . . . the second formal principle: *If an interpretation is to be accepted as a valid description of the reality the data are referring to, it must be the only one that gives a complete and reasonable explanation of the information available.* [italics in the original] (p. 170)

Ethnographic Methods

Ethnography (or ethnomethodology) is a strategy for studying the commonsense features of everyday situations—the common, ordinary happenings in a particular setting of interest. In these studies, social interaction as an ongoing process is scrutinized and recorded in descriptive detail (Fetterman, 1989; Trueba 1991; Trueba et al., 1990).

Ethnography, whose roots are in anthropology, has about a 30-year history of being seriously utilized in adapted ways in education. In the early 1960s, Stanley Diamond at the New School for Social Research in New York City established a series of conferences on anthropology and education. This program, called the Culture of Schools Program (Wax et al., 1971), was funded by the U.S. Office of Education (USOE). In 1965, the program was moved to the sponsorship of the American Anthropological Association, with USOE support intact. This, then, evolved into the Program in Anthropology and Education (PAE), subsequently directed by Fred Gearing, who, with M. L. Wax, developed ongoing conferences. In 1970, the Council on Anthropology and Education formally organized and began a newsletter.

While the cultural aspects of educational activities are now investigated through anthropologically based methods, the traditional training one receives in education research is not comparable to the six-year training an anthropological researcher typically receives. A research training seminar entitled Anthropology Field Methods in the Study of Education was offered for the first time at the American Educational Research Association's annual meeting in 1968 (Sindell, 1969).

Khleif (1974) details several key differences between *traditional* anthropological field research and *school-based* field research. These differences serve as cautions to those in education who falsely assume that the methods of ethnography can be simply transferred to the educational site while keeping its assumptions intact. (See Fetterman, 1989, and Trueba, 1991, for a detailed description of research methods of cultural ethnography.)

First of all, unlike the typical anthropological research in which culture shock is one stage of the field study, no such phase is encountered when entering the familiar classroom or school. Khleif (1974) explains:

Culture shock, as the *sine qua non* of fieldwork in exotic settings, is a learning experience; its product is the enlargement of awareness. During culture shock, the fieldworker is culturally debriefed and deprogrammed; his frame of reference is reshuffled; the habitual cultural straitjacket into which he had been imprisoned is loosened. (p. 390)

Second, the school-based ethnographer does not become a participant. Neither the teacher role nor the pupil role is assumed. Referring to Claude Lévi-Strauss,

to understand structure in human organization, the fieldworker needs to take the role of the outsider, the observer; to understand process or how a social system is maintained or slowly changed, the fieldworker must go native, that is, be an insider, a participant. This means that, at best, the fieldworker in public schools can understand only structure, not process, and thus remains more of a stranger than a friend. (Khleif, 1969, cited in Spindler, 1974, p. 391)

Khleif characterizes the anthropological field-worker as ever a stranger who can never feel at home, meaning that an ethnography of schooling in our culture can only be developed by viewing the familiar as unfamiliar. He suggests that the first way to achieve this goal is systematically to use the public school as a study site for the larger society. The second way (since non-American cultures have served ethnographers, in a sense, as testing grounds for American societal presuppositions) is for school-based ethnographers to increase their insights by a process of reversal, whereby they apply concepts of non-literate cultures to literate Americans, perhaps doing so in the school setting. The third way is to see school-based ethnographers as studying the tribalization and detribalization processes in American society within the societal and cultural boundaries of schools. American customs founded in basic religious, ethnic, and occupational stratifications can be examined in the school setting as well (Khleif, 1974).

The data-collection techniques within ethnography consist primarily of participant observation, along with the strategies used in case studies and grounded theory. Spindler (1974) calls participant-observer methodology "the keystone of an anthropological approach" (p. 385). Therefore, the limitations on validity are those that have been discussed for those methods.

Many would maintain that validity is the major strength of these methods. LeCompte and Goetz (1982), in particular, point out that there are four reasons for the high internal validity of these strategies, while external validity, the degree of generalizability, is neglected. Regarding the former point, internal validity comes from the following methodological features:

1. Ethnographers commonly act as participant observers and live among the subjects over long periods of time; they are able to continually refine their interpretations over time and compare them to "reality."
2. Interviewing informants involves using phrasing and vocabu-

lary more closely in tune with the subjects' own and less abstractly than in instruments used in quantitative studies. This therefore increases the likelihood of the instrument being able to tap the information for which it was developed.

3. Participant observation is conducted in natural settings that are the reality of the life experiences of subjects more so than are contrived settings of quantitative studies.

4. The analysis in ethnography uses a process of "researcher self-monitoring," a "disciplined subjectivity" that brings the study under continual questioning. (p. 43)

There are three reasons that ethnographers ignore external validity, according to LeCompte and Goetz (1982) and Goetz and LeCompte (1984). First, the purpose is to describe in detail aspects of a single subject, group, or unit. And, even if multiple sites are used, the researcher is obligated to enter each site as if he or she had no other information and as if this site were unique. Generalizability, then, is precluded. Second, the ethnographer enters the field site without assumptions, preconceived notions, or hypotheses. Therefore, there are no bases for comparison or generalizability. Third, the problem studied, the nature of the goals, and the application of the findings differ substantially from traditional quantitative methods, and so definitions of external validity must vary.

Concerning the problem studied, the credibility of quantitative designs is based on examining effects in *controlled* situations, looking at variables uniquely, one at a time. In contrast, according to the discussion in LeCompte and Goetz (1982), ethnographers focus on the "interplay of variables in natural context." Credibility is based on examining "all causal and consequential factors" (p. 33). Regarding the goals of studies, the goal of ethnographic research is to develop theory not to test it, which requires that a priori relationships be avoided. While quantitative researchers aim to generalize from the sample to the population, and external validity must (by definition) be built in, ethnographers aim for comparability and translatability in order to apply their results. Comparability differs from generalizability and "requires that the ethnographer delineate the characteristics of the group studied or constructs generated so clearly that they can serve as a basis for comparison with other like and unlike groups" (p. 34). Translatability is the aim—the quality that "assumes that research methods, analytic categories, and characteristics of the phenomena and groups are identified so explicitly that comparisons can be conducted confidently" (p. 34).

Phenomenological Research

Not too different from the methods described thus far in this section is phenomenological research. As described by Yvonna Lincoln, based on the premise that one socially constructs reality, the phenomenologist "looks in natural contexts for the ways in which individuals and groups make sense of their worlds" (1990, p. 290). The phenomenological researcher "collects" the realities of the participants and the interpretations of their constructions. Lincoln distinguishes this researcher's task from the quantitative researcher's by its expansionist purpose; it is contrary to the numerical reductionism of the quantitative researcher's work. Van Manen (1990) describes this as "human science." He explains:

> From a phenomenological point of view, to do research is always to question the way we experience the world, to want to know the world in which we live as human beings. And since to know the world is profoundly to be in the world in a certain way, the act of researching—questioning—theorizing is the intentional act of attaching ourselves to the world, to become more fully part of it, or better, to become the world. (p. 5)

Kvale (1983) describes three aspects of this method: open description, investigation of essences, and phenomenological reduction. The first step, *open description*, is for the subject merely to describe directly his or her experience as completely as possible, extemporaneously, with no consideration of cause or origin. The second step, *investigation of essences*, involves, as Kvale puts it, "varying a given phenomena [*sic*] freely in its possible forms, and that which remains constant through the different variations is the essence of the phenomenon" (p. 184). Finally, *phenomenological reduction* involves "suspension of judgment as to the existence or non-existence of the content of the experience" (p. 184). Simply stated, in this particular phase of the method (sometimes called *bracketing*), the researcher puts in parentheses his or her foreknowledge and common sense about the emerging phenomenon to help arrive at the essence of it. To relate it to quantitative research philosophy, this operation is a way of accounting for the researcher's prior knowledge, subjective biases, and expectations. Kvale puts it this way: "The phenomenological reduction does not involve an absence of presuppositions, but a consciousness of one's own presuppositions" (p. 185).

Dale Howard (1994), using a phenomenological research approach to study the meaning of adults confronting computer technology for the first time, states,

> This, then, is a phenomenological investigation of the experiences of those persons who have sat down to the keyboard and can articulate the experience of being introduced to a computer. Another interpretation of the "text" may help us "see" computer technology as if for the first time. (p. 34)

Even though phenomenology is private and holistic, it can be defended as *not* antiscience. Because human experience is unique, one cannot detach and reduce external data; further, Kvale (1983) implies, it is this inability to abstract that forms the existential nature of psychology. That is, general laws and theories cannot be applied to a specific individual in a unique set of circumstances. Others, however, argue that emerging sets of themes from many subjects may, in fact, form the essence of a generalization applicable to those in similar states of life.

In an in-depth phenomenological study of educational leadership, Mitchell (1990) explains that he selected this approach to inquiry to get at the "lived experience" of educational leaders. Based on the original ideas of Edmund Husserl 85 years ago, Mitchell assumes that philosophy can be independent of objective science and that "the ultimate foundation of all knowledge is in human consciousness" (p. 255). Because phenomenology is, in fact, not like the scientific method, a unified system or set of methods that can be learned, it is "more a method of study, a way of viewing, a perspective, a stance, a manner of thinking . . . and its basic tool is 'seeing' and 'interpreting' what is seen" (p. 253). Mitchell cautions that researchers who apply the phenomenological approach in education are not all engaged in the same pursuits.

Appearing in Appendix B is a summary of a study on humor by Foerstner, Newman, and Koenig (1985) that demonstrates one method of phenomenological research. In this example, the subjects are allowed to discuss humor freely and in their own words, while they alone structure their responses. In the analysis, the transcription is carefully reviewed while attempting to maintain maximum openness. Following the first reading of the responses, the central meaning units expressed by the subjects are explicated. Then they are related to the whole to get at their central themes, their essence.

Enhancing Validity of Phenomenological Methods

Wertz (1986) makes the claim that, for reliability in phenomenological research (the process of defining essential themes from informant experiences) the themes must be present in every informant's experience. They are "invariant despite changes in the factual details" (p. 197).

One method of enhancing validity that crosses many other methods is triangulation. *Triangulation* is the combination of several data-collection strategies or data sources in the same design. Jick (1979) traces the concept of triangulation in social science to what Campbell and Fiske (1959) call *multimethod-multitrait research strategies*. Its use in the social sciences stems from the concept of triangulation in the military. Simply put, in navigation strategy, using multiple reference points (and not just three) enables one more easily to pinpoint an object's exact position (Denzin, 1988).

From Denzin's (1978a) work, Jick (1979) describes the between-methods type and the within-methods type of triangulation. The between-methods type he describes as

a vehicle for cross validation when two or more distinct methods are found to be congruent and yield comparable data. For organizational researchers, this would involve the use of multiple methods to examine the same dimension of a research problem. For example, the effectiveness of a leader may be studied by interviewing the leader, observing his or her behavior, and evaluating performance records. (p. 602)

Between-methods triangulation is the most conventional form and is used to test the data for the degree of external validity. Its frequent use is based on this fundamental assumption: "The effectiveness of triangulation rests on the premise that the weaknesses in each single method will be compensated for by the counter-balancing strengths of another" (p. 604). And, for all practical purposes, this type can be assumed to be the standard usage of triangulation. The second type, within-methods triangulation, is rare. It deals with reliability and "essentially involves cross-checking for internal consistency " (p. 603).

This form of triangulation is much less frequently used because it is limited to the use of just one method.

Essentially, one would use several variations of one method to collect several sets of data, which would then be compared. Jick (1979) reviews the advantages and disadvantages of triangulation. First, the advantages:

1. Allows researchers to be more confident of results
2. Can stimulate creative methods, new ways to "capture" a problem
3. Can help "uncover the deviant or off-quadrant dimension of a phenomenon"
4. Can lead to enriched explanations of research problems
5. Can lead to a synthesis or integration of theories
6. Can serve as a test of competing theories (because of its comprehensiveness)

Second, the disadvantages:

1. Replication very difficult, if not impossible
2. Is of no use if the wrong question is being asked; "if the research is not clearly focused theoretically and conceptually, all the methods in the world will not produce a satisfactory outcome" (p. 609)
3. Should not be used "to legitimate a dominant, personally preferred method. . . . if either quantitative or qualitative methods become mere window dressing for each other, then the design is inadequate or biased" (p. 609)
4. Must justify the use of the multiple methods (e.g., cannot assume all are equally sensitive to the phenomenon being measured)
5. May not be suitable for all research purposes

6. Demands creativity

7. Is expensive in terms of both time and cost

Using multiple sites is one technique suggested by LeCompte and Goetz (1982) to increase validity. Admittedly, though, the examples they provide do not meet the standard for generalizability that quantitative assumptions would require. The extent to which four factors—selection effects, setting effects, history effects, and construct effects—are present reflects the increased validity of the study, according to LeCompte and Goetz. *Selection effects* simply force the ethnographer to compare only constructs among groups where they occur. The first issue for the researcher is to match the phenomena under study with the nature of the groups. If not carefully done, one might begin a study under inaccurate assumptions regarding the nature of the groups at the various sites. *Setting effects*, too, can be diminished. An example given by the writers is that using teachers as observers in classrooms resulted in a teacher-classroom interaction that made their data characteristically different from the classroom-observation data of nonteachers. To help counteract this effect, LeCompte and Goetz suggest that one have both a teacher and a nonteacher observe and report. Setting effects also occur when groups are frequently under study (such as happens when schools are near universities), which can be counteracted by choosing nonresearched groups. *History effects*, when counteracted, will increase validity. When using more than one site in an ethnography, the various historical foundations of those sites need to be acknowledged. Three nursery schools may be studied in detail using ethnographic methods but the development of each may have evolved from extremely different roots. *Construct effects* occur under several conditions, according to LeCompte and Goetz. First, when constructs under study are idiosyncratic to groups under study, appropriate comparisons to the other groups diminish. That is, the ability to make appropriate comparisons to other groups is lessened. Second, to the extent that the use of any observational instruments is not common across groups, there are likely to be construct effects. Third, the meaning of phenomena might vary across groups, creating construct effects.

Conclusion

By attending to the 14 qualitative methods, one is more likely to become sensitive to the design validity of a study. The more evidence of design validity a study can show, the more truth value we can presume. Similarly, these concerns are revealed in quantitative studies by the application of concepts of internal and external validity (Campbell & Stanley, 1963). With a predominantly quantitative study, the more internal and external validity a study has, the more confident we are in the truth value of the study.

A good researcher needs to be familiar with a variety of methods. Multiple methods may enhance the quality of a research study. Lombard (1991) demonstrates how the use of multiple methods increased the quality of her research. In her case, however, all of the methods she uses tend to be qualitative: interviews, member checks, and critical-incident techniques. These techniques are described by J. T. Bailey (1956) and by Placek and Dobbs (1988). Stivers and Srinivasan (1991) also use multiple methods to improve their insights and the interpretability of research, and they use mathematical algorithms and mathematical simulations to enhance and support their interpretations. Increasingly, researchers are using multiple methods to improve the quality of their research. Doing so is consistent with the qualitative-quantitative continuum, which is based upon the assumption that investigators should not be tied to any single methodology. The research question always should dictate the method.

5

Applying the Qualitative-Quantitative
Interactive Continuum to
a Variety of Studies

WE BEGIN THIS CHAPTER with procedures we suggest can be used in critiquing research. Then we offer actual examples of research published in the disciplines of education and counseling, contextualizing them within the continuum so that their validity can be evaluated.

The process of critiquing each study involves assessing the methods the researchers use. The methods, you will recall, we present in Figure 3. The totality of the "methods" we call the *design*. When critiquing a published study, one is limited to knowing only what is written in the article about the methods the researcher uses. Full accounting for each activity on the part of the researcher may or may not be included. Our judgments about each of these studies are limited, therefore, as are all critiques of published work.

More important than the conclusions we draw about these four studies is the process we suggest. Others may ask somewhat different questions about each study. We are not particularly bothered by that. In fact, our questions here are not uniform from study to study. The bottom line for us is advocating for a critique of published research that seeks to judge whether the research question is consistent with the research methods. Our process is only one of several that can accomplish that goal.

After reading chapter 5, the reader should be able to

1. Discuss why utilizing the continuum increases the researcher's awareness that research is a holistic endeavor
2. Pose and answer consistency questions across a research study in the areas of purpose and methods, methods and data, purpose and conclusions, and implications and purpose
3. Apply the continuum to a published research study
4. Apply the continuum in planning a research study

Step-by-Step Procedures to Use in Critiquing Research

In evaluating research studies, the researcher can apply the continuum as an assessment scheme. In planning a research study, the researcher can utilize the continuum to assess his or her plans. And, because research is conceptualized as an unbroken continuum, one may enter the continuum and make inquiries for assessment and critiquing purposes at any point. The following steps are a convenient place to begin, especially when one is initiating the research process. In the *consistency-questions model*, one asks,

a. What is the question, purpose, or reason for doing the research?
b. What research methods might one use to address this question, purpose, or reason?
c. Contingent on the answer to question b, what are the underlying assumptions of the research method defined?
d. What are the findings of the research?
e. What are the implications of the findings of the research?

As illustrated in Figure 4, the sequence of questions forms a loop, implying continuity and consistency. Because of this mapping, the researcher is able to assess the consistency from any point in the loop

88

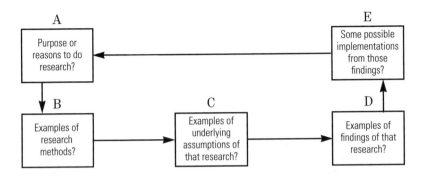

Figure 4. Does the research question dictate the research method?
Model of the consistency questions to ask in critiquing research

to the adjacent point: Is there a match between the question or purpose (*A*) and the methods (*B*)? Is there a match between the methods
(*B*) and the assumptions (*C*)? Is there a match between the implications (*E*) and the purpose of the question (*A*)?

For example, one might carry out research, as in the early school
effectiveness studies (Olson, 1986), in which the purpose is to explore
what characterizes a school where learning gains are relatively high.
To acquire thick descriptive detail, this question dictates the use of
qualitative methodology. Following such a study, however, the investigator should not generalize from the descriptions to other schools. Generalization is consistent neither with the purpose of the study nor with
the underlying assumptions of the specific research methods. If one
were to generalize, the generalization would be criticized as inappropriate and in violation of the assumptions.

This consistency-question approach as depicted in Figure 4 is a subset of the interactive continuum represented in Figures 1 and 2. The
continuum assumes that the research question dictates the methodology. If the researcher uses methods consistent with his or her purpose, the conclusions likely will be consistent with the underlying
assumptions of those methods. The consistency questions to ask are
common to all research studies.

Figures 5 and 6 illustrate possible answers to the consistency questions when either the qualitative or the quantitative paradigm is pre-eminent. The two figures would ideally be depicted in one schematic drawing, which would be more in keeping with the continuum. They have been separated here for illustrative purposes. For example, if a researcher asked question *A* in Figure 5 or 6 ("What is the research question?"), the answer (that the researcher wanted to test a set of hypothesized relationships) would lead to the point on the continuum illustrated in Figure 1 corresponding to that purpose, square 3. Examining Figure 1 at square 3 shows that testing hypotheses is derived from a review of the literature, square 2, and is followed by collecting the data, square 4. The researcher who has defined his or her purpose as describing a certain phenomenon in detail, with no preconceived hunches or hypotheses, would enter the research continuum in Figure 1 at circle *A*. As is apparent in Figure 1, this is the first step in utilizing the qualitative part of the continuum. It is followed by analyzing the data (circle *B*), drawing conclusions (circle *C*), attempting to derive hypotheses (circle *D*), and perhaps developing a theory (circle *E*), which places the research within the area of overlap with the quantitative part of the continuum. Using the continuum forces the investigator to perceive the research in a holistic context, in a context of both qualitative and quantitative assumptions rather than within a narrow bias of either one or the other.

The consistency questions (Figure 4) assist the researcher in ensuring that consistency exists throughout the research, from questions or purpose through implications. In Figures 5 and 6 the questions are placed so that consistency is maintained throughout both paradigms (along the continuum). Accepting the continuum implies accepting that, consistent with the central place of theory, all other components must coexist in appropriate relationship to it. Adopting the model and planning research within its structure permits research to be carried out consistently and, thus, with optimal validity. Researchers who adopt

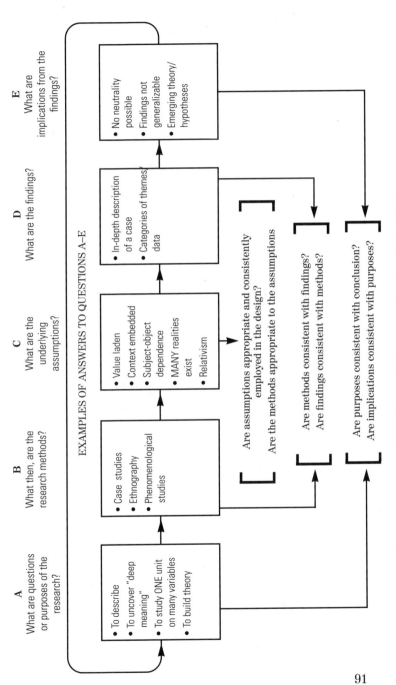

Figure 5. Model of the consistency questions to ask in critiquing research: The qualitative paradigm being dominant

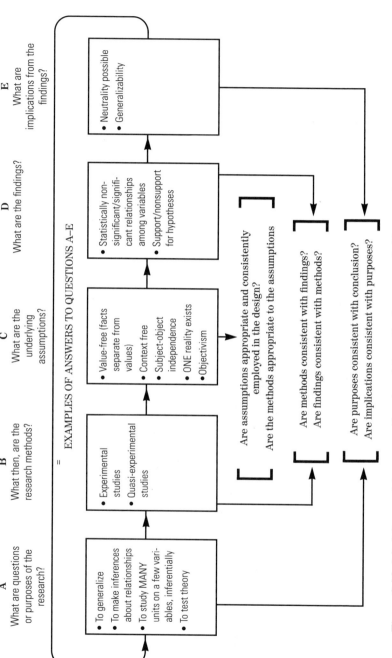

Figure 6. Model of the consistency questions to ask in critiquing research: The quantitative paradigm being dominant

the model assume a philosophy of research that is unified and focuses on question and method consistency not on ideology.

To conceptualize the logic in applying the interactive continuum, we suggest three phases that flow naturally from what has just been presented. In the first phase, the continuum assumes consistency between question and method. In the second phase, one evaluates the extent to which there is consistency. And in the third phase, each study can be examined closely for issues of design validity.

In the next four sections, we present suggestions for applying the continuum and its subset of consistency questions; then we present suggestions for assessing issues of design validity to evaluate research. Four published studies are critiqued (Appendixes C–F). We want to emphasize that the process being suggested can be generalized to virtually any situation. It can be adapted, used flexibly. We emphasize the *process* of critique—asking the questions about the research studies one reads is the important activity. Two people may come to different judgments about the truth value of any one article. Each critique is a value judgment. The four articles we refer to here may be critiqued by others using our model, but their conclusions might differ in some limited ways.

These critiques exemplify the second and third phases of research evaluation, that is, examining issues of design validity as defined at the end of chapter 3. We use our model to critique each article and then add more general reflections after each critique. The reader will notice that the four critiques, while following our model conceptually, vary widely in how they are carried out. We are less concerned with the specifics than with the value of such critiques and that they are actually conducted.

Critique of the Alexander and Harman Study

The first study to be critiqued, "One Counselor's Intervention in the Aftermath of a Middle School Student's Suicide: A Case Study," by Alexander and Harman (1988), can be found in Appendix C. The reader

93

will find the critique more meaningful if the article is read before continuing.

The purpose of this study is to describe an intervention for middle school students who were affected by a classmate's suicide. There is no statistical analysis, and the authors indicate that attempts to generalize should be avoided. This, then, clearly falls into the broad category of qualitative research, and we analyze it in terms of the qualitative research components previously identified.

Neutrality

In terms of neutrality, this study is weak because there is only one observer, and there was no apparent attempt to control for personal bias.

Prolonged Engagement On-site

Prolonged engagement on-site seems to be sufficient for this study because the counselor was on-site prior to the suicide and counseled some students for up to 12 weeks following the incident.

Persistent Observation

There is no way to assess the students' behavior before, as compared to after, the suicide. The inference of this study is that there was an increase in suicide ideation and a fear of additional suicide attempts. No data are presented to substantiate this claim.

Peer Debriefing

According to the study, there was no attempt at peer debriefing. No other counselors were sought to check personal perceptions or bias or to offer other interpretations.

Triangulation

There was no attempt to obtain information from other sources.

Member Checking

As a result of the counselor's interaction with the students, she concluded that certain themes emerged, such as poor self-concept and excessive self-demands. There is no report of checking these themes for persistence in or consistency among group members.

Referential Materials

There is no way to assess whether referential materials were used—none are identified, and we can only assume none could be made available.

Structural Relationships

No other data are shown to have been available to interweave for the purpose of establishing structural relationships between data sets. She does, however, attempt to interweave some Gestalt theory into her conceptualizations.

Theoretical Sampling

No attempt at theoretical sampling is described.

Leaving an Audit Trail

There is no indication that notes, records, or any other kind of documentation were kept. It appears to be entirely personal observation.

Related to Generalizability

Applicability. To estimate the applicability of this study, one needs deep descriptors to clearly define the characteristics of the sample. Description is not of sufficient detail to have a clear sense of socioeconomic status, culture, and so on.

Context limited. There is no indication that the interpretation is context free. On the contrary, there is a good possibility that it is context specific. The concepts that the authors apply are basically Gestalt, and the interpretations are all from this perspective, as is the investigator's training and predisposition.

Replicability. No data are available to estimate the consistency of the findings.

Negative Case Analysis

No evidence is available.

Truth Value

The purpose of this study is to identify the potential effectiveness of a Gestalt approach in helping children who may be suffering from the trauma of a classmate's suicide. In light of the above characteristics, or validity criteria, used in evaluating qualitative research, we have limited information and are skeptical about the truth value of this study as it relates to making a strong statement that Gestalt therapy is a viable intervention approach in these situations.

Reflections

By reflecting on the Alexander and Harman (1988) study through the lens of the interactive continuum, one can gain insights into both the strengths and the limitations of its design validity. From these insights, future researchers interested in replication, when making decisions about methods, might want to consider what has been revealed.

This article is appealing in that it is well written, intuitive, and deals with some obvious truths about the need to be sensitive to children who have experienced a suicide. However, it is methodologically weak, at least insofar as what is reported in the published article. Reporting more thoroughly on all aspects of methods and improving on those that are omitted would strengthen its potential impact on knowledge about intervention strategies. Future researchers may want to take these considerations into account when designing a study.

Given what we know about the study, there are techniques that could have enhanced the validity of this study. Alexander could have enlisted the aid of other observers, which would have helped counterbalance the effects of personal bias and sensory deficit. Checking her interpretations with experts in the field also could have helped make the study more valid; and the use of different forms of data collection to supplement the unstructured interviews would have been helpful.

If one applies the interactive continuum of qualitative-quantitative methodology, the themes that emerge from this case study are clearly qualitative, yet these themes provide feedback appropriate to the quantitative end of the continuum. The themes could become the basis for a study more empirical in design. Students could be randomly assigned to treatment groups, such as Gestalt, behavioral, and no treatment (control, perhaps a group experience based on a self-help model without formal theoretical basis). The themes that emerge from the case study could then be used to evaluate treatment effectiveness, while controlling for such variables as age and sex to increase the ability of the tests to determine any effect due to treatment alone. Alexander and Harman's mix of qualitative and quantitative characteristics is not in itself a problem, but their study might have been enhanced had they elaborated upon this mixture and had they submitted variables developed by the case study to an empirical investigation. A strength of this study is that it sensitizes people to an important topic. The research also may have heuristic value in that it may lead to further study. However, before one could conclude that Gestalt therapy is effective, more control in the research procedures would be necessary.

Critique of the Curtis Study

The second study to which we apply the continuum is "Effect of Therapist's Self-Disclosure on Patients' Impressions of Empathy, Competence, and Trust in an Analogue of a Psychotherapeutic Interaction" by John Curtis (see Appendix D).

The critique of the Curtis (1981) study is a composite done by students in a Ph.D. counseling research seminar.[1] While similar to the other critiques, there are some real differences. Different evaluative questions are used, for example. The students analyzed the study by describing (1) the research question; (2) the statistical questions, or hypotheses; (3) the research design in Campbell and Stanley's (1963) terms; (4) the relationship between the research question and the design; (5) the statistical model; (6) the conclusions of the study; (7) the recommendations that the students would suggest; as well as (8) general quantitative considerations; and (9) general qualitative considerations. We include this critique to show that it is the process of evaluating a published research article that is important; the specific format one uses is less important.

Research Question

Is there a relationship between a therapist's self-disclosure and the patients' perceptions of the therapist's empathy, competence, and trust?

Statistical Questions

1. High therapist self-disclosure condition will yield patients' lowest evaluations of empathy, competence, and trust

2. Low therapist self-disclosure conditions will yield patients' moderate evaluations of empathy, competence, and trust

3. No therapist self-disclosure conditions will yield patients' highest evaluations of empathy, competence, and trust

Research Design

Refer to symbols described in chapter 3.

$R \quad X_1 \quad O \qquad X_1 =$ high disclosure
$R \quad X_2 \quad O \qquad X_2 =$ low disclosure
$R \quad X_3 \quad O \qquad X_3 =$ no disclosure

THREATS TO INTERNAL VALIDITY

Threats from history, maturation, mortality, instrumentation, testing, and statistical regression were controlled for. Selection bias was uncontrolled, as volunteers who were ongoing clients agreed to participate in the study. The author fails to explain randomization procedures.

EXTERNAL VALIDITY

1. Reactive effects

   a. Re-posttesting—no problem indicated

   b. Subjects aware of participating in research; could have had reactive effect, creating bias in responses (Hawthorne effect)

   c. Rosenthal effect (effects caused by investigator's behavior or appearance) could be a factor, but actual dialogues of experimenters not provided by author

2. Generalization to population: outcomes can only be generalized to people who meet five criteria indicated in study

3. Generalization to other settings: dependent and independent variables not sufficiently defined by author; no sample provided of independent variable

CONSTRUCT VALIDITY

1. Author's definition of self-disclosure defined narrowly and not necessarily consistent with other theorists' views of construct

2. Author's operational definition of self-disclosure not consistent with psychodynamic view, which is assumed by investigator

3. Operational definitions of constructs measured by test instruments not defined

4. Two separate measures of empathy used without defining construct of empathy or how each instrument defines empathy

Relationship Between Research Question and Design

The results of the study are highly misleading regarding therapist self-disclosure. In effect, the research design does *not* reflect the research question; that is, self-disclosure is not defined accurately.

Statistical Model

The author uses three one-way ANOVAs without indicating alpha correction. Alternatively, he could have used the general linear model (MANOVA, or factor analysis).

Statistical Answer/Results and Conclusions

1. Subjects' highest evaluations of Perceived Empathy I occur in Dialogue III (no self-disclosure)

2. Subjects' highest evaluations of Perceived Empathy II occur in Dialogue III (no self-disclosure)

3. Subjects' highest evaluations of Perceived Competence occur in Dialogue III (no self-disclosure)

4. Subjects' highest evaluations of Perceived Trust occur in Dialogue III (no self-disclosure)

Curtis (1981) concludes that there is some doubt about the effectiveness of a therapist's self-disclosure as a therapeutic technique. Additional research is needed to assess the effectiveness of self-disclosure in relation to other theoretical models to see if results would be consistent with those found in this study.

Recommendations

1. Use alternative method of statistical analysis
2. Include better operationalization of constructs; instruments presented with limited reliability statements (and as reference to validity factors)
3. Include procedures for random selection and assignment

Quantitative Considerations

Randomly assign clients to therapists who work under the three conditions (high, low, and no self-disclosure). Assess the clients' perceptions of the dependent variables, empathy, trust, and competence.

Qualitative Considerations

The following questions emerge from Curtis (1981) and could drive related qualitative studies:

1. What type of therapeutic outcome would the subjects project, based upon the vignette read (level of self-disclosure)?
2. Why did the subjects think they were being assessed? What did they think was the researcher's intent?
3. What were the subjects' general impressions in response to the therapist in the vignette? What kinds of reactions did they have?
4. Based upon previous counseling experiences with a therapist's level of self-disclosure, what were the subjects' experiences and expectations of counseling?

Curtis (1981) himself speculates about possible qualitative studies grounded in this research.

Reflections

One may enter the qualitative-quantitative continuum from either a qualitative or a quantitative perspective (refer to Figures 1 and 2). One may then critique the study under consideration and suggest future qualitative or quantitative research (symbolically represented by the feedback loops in Figure 2). This student critique demonstrates one asset of our model—it has heuristic value: any particular study can lead to other questions and other research.

Critique of the Fuller Study

The third study to which we apply the continuum is "The Monocultural Graduate in the Multicultural Environment: A Challenge for Teacher Educators," by Mary Lou Fuller (1994), found in Appendix E.

Neutrality

That the 28 participants were recent graduates from one university (multiple cases, similar time of graduation, and a common university experience) would seem to help the researcher diminish bias and build neutrality. Taped interviews and observations were transcribed, lending reduced personal bias to the data as they were analyzed. But the researcher who collected the data also analyzed the data, rendering personal bias not completely eliminated.

Prolonged Engagement On-site

The goal of prolonged engagement as a design feature is to take into account distortions in typical experiences of those being studied. The original conceptualization of prolonged engagement grew out of ethnographic studies. The researcher was alerted to such things as being a "stranger in a strange land" (Lincoln & Guba, 1985, p. 302), selective perception, and building trust in the research setting. Fuller's (1994) study, while qualitative, was not an ethnographic study. The researcher engaged in 28 settings rather than one; therefore, the notion of prolonged engagement does not strictly apply.

Conceptually, however, one might conclude that interviewing and engaging in observation of 28 teachers to study the multiple ways individual teachers interact with different cultures is more in keeping with the goal of exploring what is typical (undistorted) in teachers' experiences than including only 1 or 2 teachers. In other words, to the extent that 28 provided more evidence for common meaning from research subjects as well as for clarifying and decreasing the potential biases in the researcher's perceptions, this study was strengthened.

Persistent Observation

The object of persistent observation as a design feature is to achieve depth of meaning from the data (i.e., what seems salient in the setting). Like the other criteria, it was originally described for ethnographic research. To comply with this criterion, the researcher focuses in detail on the most relevant factors in an ethnographic study. The emerging domains of meaning, then, are based on a depth of understanding. To apply this characteristic to the Fuller study (not an ethnographic study) requires examining how the researcher determined what labels to apply to the emerging themes of the teachers' experiences. In this instance, the salience of the themes—that they went beyond superficialities (Lincoln and Guba, 1985)—was substantiated by including within the thematic interpretations only those ideas expressed by at least one-third of the 28 teachers.

Peer Debriefing

No evidence was available.

Triangulation

Because two data-collection methods (observation and interview) were used, triangulation is strengthened. No information is given about the consistency of the data collected between the two techniques. Multiple sources of experience (28 teachers) is additional evidence of triangulation.

Member Checking

No evidence was available.

Referential Materials

No evidence was available.

Structural Relationships

There is no evidence that the resulting data from the two methods (observation and interview) were compared for structural relationships.

Theoretical Sampling

No evidence was available.

Leaving an Audit Trail

No evidence was available.

Related to Generalizability

The author indicates what is needed for preservice, monocultural teachers who will be teaching in multicultural settings, thus showing her intent to generalize.

Applicability. While the results do add to the knowledge base about multicultural environments, Fuller limited the study to the cultural environments of schools in Texas, Nevada, and Arizona. The narrative does not provide deep enough description to be able to estimate applicability. In addition, the participants in the study were self-selected. For whatever reason, each chose to teach in his or her location. Those reasons may differ for other teachers to whom the results might be applied.

Context limited. There is insufficient information to assess the extent to which the context limits the generalizability. One common feature across the 28 interviews is that the teachers may have been responding in ways that they thought might satisfy the interviewer.

Replicability. There is no way to estimate whether the results would occur again at different times or in different settings.

Negative Case Analysis

No evidence is provided in the published study that outliers or processes for including them were included in the design.

Truth Value

Truth value, we suggest, is the overall judgment based on all the preceding criteria. The study's limitations in meeting the 13 criteria suggest a moderate level of truth value.

Reflections

What can be done to add to Fuller's (1994) research from a quantitative perspective? Among many possibilities, a couple of ideas come to mind. The author identifies six categories of meaning about these teachers' multicultural experiences. A survey of items based on these themes could be developed. Ratings from a sample of preservice teachers would allow interrelationships among the themes to be analyzed. Perceptions and experiences could be measured and the extent to which the six themes are generalizable could be determined. Here, again, the interactive continuum has value in showing how qualitative research can lead to quantitative research and vice versa. Science, after all, is the ongoing accumulation of knowledge.

Critique of the Rhoades and Kratochwill Study

The fourth study to which we apply another version of the continuum is "Teacher Reactions to Behavioral Consultation: An Analysis of Language and Involvement," by Rhoades and Kratochwill (1992), found in Appendix F.

Research Question

Does the language of the behavioral consultant and the involvement of the teacher in the behavioral consultation process affect the acceptability of the intervention to be used in regular classrooms?

Independent Variables

Teacher: Involvement vs. Noninvolvement
Language: Technical vs. Ordinary

Dependent Variable

The dependent variable is the acceptability rating for each consultation scenario.

Design in Campbell and Stanley's (1963) Terms

$R \quad X_1 \quad O_1$
$R \quad X_2 \quad O_1$
$R \quad X_3 \quad O_1$
$R \quad X_4 \quad O_1$

R = random assignment

X_1 = technical language and teacher involvement

X_2 = technical language and no teacher involvement

X_3 = nontechnical language and teacher involvement

X_4 = nontechnical language and no teacher involvement

O_1 = acceptability rating

Internal Validity

The strengths of this study are that it is a true experimental design and has random assignment of teachers to the four groups. Theoretically, then, this design controls for all threats to internal validity: history, maturation, testing, instrumentation, regression, selection, mortality, and interaction of selection and maturation.

External Validity

There are two threats to external validity in the Rhoades and Kratochwill (1992) study: volunteers were used; and there may be reactive arrangements because of the video. The major problem with this design, however, is that it is analogue; that is, it is a simulated design. People respond to videotapes but we do not know if they would respond the same way with actual encounters. In addition, the subjects were only exposed to 12 minutes of video. In a real-life situation, the exposure would be longer.

Conclusions

In the article, the authors conclude that there are no main effects for involvement and no main effects for language. They should have said that there are no significant main effects. Their stated conclusion implies the acceptance of the null hypothesis, which should not be done.

Qualitative Research Suggestions

As the authors suggest, interviewing the subjects after they watch the videotapes could be done in a qualitative study to identify why some people responded as they did and thereby suggest what needs to be done quantitatively in future studies to make the results more generalizable.

Reflections

It is obvious that this critique is less exhaustive than the critique of Curtis (1981), the other quantitative study. We intended this abbreviated version to show that, even at a broad and general level, use of our continuum model can lend valuable insight into the quality of research.

6

Conclusions: Modern-Day Science Is Both Qualitative and Quantitative

SHULMAN, IN THE INTRODUCTORY CHAPTER to the third edition of *The Handbook of Research on Teaching* (1986), urges support for the creativity embodied in research that ventures beyond the rigid confines of typical quantitative and qualitative camps. But he voices warnings as well.

> I will therefore argue that a healthy current trend is the emergence of more complex research designs . . . that include concern for a wide range of determinants influencing teaching practice. . . . These "hybrid" designs, which mix experiment with ethnography, multiple regression with multiple case studies, process-product designs with analyses of student mediation, surveys with personal diaries, are exciting new developments in the study of teaching. But they present serious dangers as well. They can become utter chaos if not informed by an understanding of the types of knowledge produced by these different approaches. (p. 4)

When one always begins with the research question as the dictator of the method, the danger Shulman warns against is almost completely eliminated. In this concluding chapter, we emphasize that the conceptualization of the qualitative-quantitative interactive continuum maximizes the validity, reliability, and utility of the "exciting new developments" on the methodological horizon.

One underlying assumption of this book is that any interpretation of data is only as good as the accuracy of those data. If the data on which researchers are basing judgments are faulty, one cannot expect the conclusions to be accurate. Obviously, one can have appropriate inferences and conclusions without collecting data. Einstein knew his theory of relativity was correct before he had the data to support it. He spent most of his life collecting data to support the accuracy and usefulness of his theory because he, too, probably accepted the assumption that one can only have confidence in a research theory to the degree that one has reliable and valid data that support the theory. Therefore, high confidence is achieved only if judgments are based upon accurate data.

It follows, then, that one key to good research is the accuracy of the data-collection procedures (which are the concepts of reliability and validity) regardless of the label given the analysis, qualitative or quantitative. This theme is consistent with the work of Goetz and LeCompte (1984) and with the basic underlying assumptions of the modern-day philosophy of scientific inquiry. Therefore, in considering our thesis, regardless of whether the discussion is of qualitative or quantitative methods or of the interaction between them, there is a recurring attempt to make the reader sensitive to the questions that one must ask about the data: reliability and validity. We believe that consumers of research profit most by asking appropriate questions about the accuracy of data-collection procedures before accepting any researcher's interpretation of his or her findings.

Based on these assumptions, qualitative and quantitative research as an interactive continuum, shown in Figures 1 and 2, becomes more evident. What dictates the categorization of the method is the type of question being asked, the type of data being used, and the selection of data-analysis techniques. Some types of data are more difficult to quantify and, therefore, are not quantified; while other types of data are not initially quantified but are quantified at a later point. Then they may be used to support or not support the qualitative analysis and/or indicate further types of research. This feedback conceptualization is represented in Figure 2.

The aim of this book is to place in perspective the relationship between qualitative and quantitative research; to elucidate methods that have tended to be classified as qualitative or quantitative, along with their respective strengths and weaknesses; and, we hope, to indicate the fluidity with which one can utilize the interactive continuum. We expect that after reading this book, the reader, when asked, "Which are better, qualitative or quantitative methods?" will not be able to answer the question. Instead, the reader would say, "It depends on the research question, the type of data, the type of analysis, as well as the point in time one is looking at the research process." By now, the reader understands our conceptualization that the modern-day scientific method of inquiry is both an inductive and deductive process, with feedback loops that affect the inductive and deductive procedures and that are self-correcting.

Furthermore, after a review of the literature dealing with the controversy between qualitative and quantitative methods, the basis for much of the heat of the debate becomes evident. Some proponents write of the process of doing qualitative and quantitative research, while others write of the underlying assumptions of qualitative and quantitative methods. The disagreement comes from the differences in conceptual levels (i.e., procedures vs. underlying philosophical assumptions). Our schematic illustration shows the underlying philosophy of the methods and demonstrates its nature as a continuum.

Attention is again directed to the methodology of the qualitative-quantitative continuum conceptualized in Figure 1. Note the philosophical differences: the starting point for the quantitative researchers (square 1) is theory, from which steps 2, 3, 4, 5, and 6 derive. Hypotheses are formed, data collected and analyzed, and conclusions drawn. This is the theory-based, deductive approach shown in sequence. On the other hand, the qualitative researchers begin with data (circle A) and proceed through steps B, C, D, and E as the data are interpreted and theory built. This is the theory-building, inductive approach shown in sequence. The dotted rectangle around the place of theory in both philosophies shows its connecting role. Conceptually, this book emphasizes that the theory is neither at the beginning nor at the end, but the quantitative (square) and the qualitative (circle) overlap on the diagram and continue the cycle, closing the gap. This conceptualization is not too far from that of Guba (1978) and Guba and Lincoln (1989) as they describe the global approach to naturalistic inquiry, depicting the sequence of such inquiry as a wave on which the researcher moves from varying degrees of a discovery mode to emphasis on a verification mode. As researchers begin an investigation, they are receptive to what might be revealed in a discovery or inductive frame of mind. Then, as patterns are revealed, the researcher moves to more controlled verifying methods in an effort to support hypotheses about the data, a deductive focus.

Debates in the literature frequently occur because researchers are arguing from different starting points, relating to the warning made by J. K. Smith (1983):

> the assumption that the two approaches are little more than alternative methodologies, whose varied employment responds simply to "what works" and not to epistemological considerations, must not be accepted at face value if we are to make the issue more intelligible. (p. 6)

Even though we feel that Smith is correct in bringing to the forefront of the dialogue the fact that researchers often make serious mistakes when they operate naively (i.e., without knowledge of the underlying assumptions of their methods), we nevertheless strongly disagree with an inference made by others that qualitative and quantitative research cannot be integrated effectively without violating the underlying assumptions of one or the other. This does not mean that one can do qualitative and quantitative research simultaneously. However, it does mean that the feedback from one approach may lead the researcher to investigate questions that arise that may require the other approach. This model is consistent with the proposition put forth by Reichardt and Cook (1979) that research needs must be the focus and that such needs may require both qualitative and quantitative methods. "Treating the method-types as incompatible obviously encourages researchers to use only one or the other when it may be a combination of the two that is best suited to research needs" (p. 11).

This interactive model resolves the gap presented by the authors of the original work on grounded theory. Explaining the void left between building and verifying theory, Glaser and Strauss (1967) pinpoint that gap, that dilemma:

> when the main emphasis is on verifying theory, there is no provision for discovering novelty, and potentially illuminating perspectives, that do emerge and might change the theory, actually are suppressed. In verification, one feels too quickly that he has the theory and now must "check it out." When generation of theory is the aim, however, one is constantly alert to emergent perspectives that will change and help develop his theory. (p. 40)

In response, we have tried to show that, within any field of inquiry, provision is made for capturing that novelty, that illuminating perspective. As these insights are generated, revised hypotheses and subsequent data collection can ascertain their stability. The claim of "no provision for discovering novelty" is canceled by this model; provision is made by use of feedback loops.

Additionally, this model may fill the void described by Miles and Huberman (1984). They present several ways of reporting the validity of qualitative data and call for a new perspective: "It looks as if the research community is groping its way painfully to new paradigms, those that will be more ecumenical and probably more congruent with the data being collected and interpreted" (p. 21). This model is consistent with their focus on the method being related to the data and to the research question.

Finally, it is important to reemphasize that one method is not necessarily better than another. It is also important when evaluating research to determine how well the research adds to the body of knowledge and to determine what it suggests needs to be done next. We believe that the model presented, the qualitative-quantitative interactive continuum, can facilitate the evaluation, planning, and conduct of research by providing a framework within which to conceptualize it.

Summary

In the early 1980s, we became concerned about researchers in the field who were jumping on paradigm bandwagons of unthought-out and simplistic biases, behaving as though there were one more effective method for conducting research than another. We responded to the growing qualitative-quantitative debate with an integrated perspective, which builds the research method on the researcher's question of interest. In a 1986 monograph, we were able to present that view to students and colleagues with whom we worked.

We have developed that perspective, taking the interactive continuum a long way from those early, preliminary speculations, not only explicating the very crucial foundational role of the research question but also expanding the interactive continuum approach to critiquing both published studies and research proposals. In Benz and Newman (1986), we generated ideas based on quantitative assumptions for increasing the validity of qualitative research. These ideas have become better developed since that time, and we have shown their congruence with the work of others.

We began this book with a philosophical base and then built our case on the notion of a false dichotomy. We took the position that the two methodologies are neither mutually exclusive (i.e., one need not commit either to one *or* to the other), nor are they interchangeable (i.e., one cannot merge methodologies with no concern for underlying assumptions). Rather, we presented them as interactive places on a methodological continuum. A researcher develops and tests theory and, as results feed back to original hypotheses, both inductive and deductive processes are operational at different points. Qualitative and quantitative methods are invoked at different points in time, and feedback loops facilitate maximizing the strengths of both methodologies.

Qualitative-Quantitative Research Methodology: Exploring the Interactive Continuum serves primarily as a practical tool and it does so for a couple of reasons. We discussed the concepts of validity and reliability and offered ways that the design validity of studies across the continuum can be improved. And we presented a step-by-step approach to evaluating research, which we hope will be helpful to consumers and planners of research.

Finally, this book is about the evolution of ideas that began a few years ago, ideas that we hope will continue to evolve. Even now, we are rethinking and expanding on some of them; so, in essence, as stated earlier, it is a work in progress. John Mitchell (1990) describes his years of contemplating educational leadership and his realization that such a study is virtually endless. We too chose to draw on our years of experience (in education, business, and consulting) to present our study in its present form, asking readers to help us clarify and develop our positions. Or better still, perhaps readers will go far beyond what we have tried to accomplish here.

Appendixes
Notes
Glossary
Bibliography
Index

Appendix A

Research Paradigms
(Based on Rist, 1977)

Quantitative	Qualitative
Analysis of causality is "component," that is, emphasis is on manipulation of variables simplifying the complexity of reality by breaking down into small component parts (i.e., variables).	Analysis of causality is "holistic." Reality cannot be broken down into components without risk of distortion; a narrow set of variables sets up a filter between researcher and phenomena under study (avoiding "closeness" to the data).
Concern is reliability.	Concern is validity.
Reliability is sought by avoiding "unreliable, biased, or opinionated data" through use "of a number of subjects."	Seek validity through "personalized, intimate understanding of the social phenomena, stressing 'close in' observations to achieve factual, reliable, and 'confirmable' data" by using an intensive case study of a very small group or some particular individuals. "It is in the interpretation of the world through the perspective of subjects that reality, meaning, and behavior are analyzed."
The world of human events is lawful; the development and verification of generalizations about that world is the first priority of the researcher.	Inner perspectives or "understanding" says truth can only be achieved by participating in the life of the observed and "gaining insights through introspection."
Knowledge is cumulative and verified through scientific methods, hypothesis testing.	Scientific method is insufficient.

Progression of knowledge is a continuum from (1) observation to (2) experimentation to (3) theory building. The use of inductive statistics stresses the experimentation-to-theory-building links rather than the observation-to-experimentation links (inductive statistics based on probability theory).

Begin not with hypotheses but with "understandings of frequently minute episodes" that are "examined for broader theory patterns."

Appendix B

Phenomenological Research: Laughter and Humor

A questionnaire was administered to a small sample of nurses and educators. These samples consisted of females between the ages of 29 and 50 years, of mixed marital status and race. Each protocol was separated into meaning units. That is, each response written by each participant was divided into a series of expressions, which, if read consecutively, match the original protocol. Next, each meaning unit was condensed to its central theme. The central themes were combined and form the final formal step in this qualitative analysis, namely, the demarcation of the typical components of "why people laugh" and "what it (laughter) does for people" for this sample of nurses and educators.

To summarize, a review of the data allows for a description of the nomothetic characteristics of the phenomena. The nursing and educator groups are united to form a single unit here because the analysis of the protocols renders any differentiation arbitrary at this exploration phase.

In describing why they laugh, the subjects in this study indicated a variety of reasons. They seemed to laugh for the sake of laughter and to enjoy life, as well as at something perceived as humorous. Laughter helped lighten stress and covered up less pleasant feelings of sadness and frustration or nervousness and shyness. While some subjects experienced laughter as an uncontrollable and spontaneous event, others highlighted their choice to laugh or how they were taught to see the brighter side of a situation. Subjects identified a capacity to laugh alone or at self, but some preferred to share laughter, especially the "contagious" type, with others.

The subjects in this study emphasized the positive benefit from laughter: they felt good, relaxed, and positive; had unity of mind and emotions; and were glad to be alive. It also provided for an emotional release to ease stress, anxiety, and tension and to cover up nervousness until subjects could regain a coping strategy. On the other hand, laughing at inappropriate times could create embarrassment, and extreme laughter could lead to a painful side and nausea (Foerstner, et al., 1985). On the following pages are a copy of the questionnaire and examples of typical components of phenomenological experience about why subjects laugh and what it (laughter) means (taken from Foerstner et al., 1985).

Results

Typical Components of Phenomenological Experience of
Why Subjects Laugh and What It (Laughter) Does for
Subjects in This Study: A Sample

I. Nurses

A. Why do you laugh?

to enjoy life, myself, and friends
allows subjects to take changes in stride
cover up feelings of sadness, frustration, inadequacy, indecision, when something is amusing
taught to see brighter side of situation, feel happy, and enjoy something
people joke, something comical happens, having fun with peers
feels good and brings joy into life if blue and down

B. What does it do for you?

feel more relaxed
eases stressful situation
like self better
easier to live with
feels good
light and airy
relieves tension and anxiety
more positive

II. Educators

   A. Why do you laugh?

      reflection of feelings, enjoy life
      at jokes, circumstances, self
      uncontrolled response to something funny
      spontaneous during conversation when a joke or quick wit of
         subjects or partner
      when a situation makes a giggle rumble inside and it bursts out

   B. What does it do for you?

      feel "up," relaxed, comfortable
      hurts side and feel nauseated if laugh very hard
      releases, feels good
      releases energy

Sample of Meaning Units and Central Themes

(A: response to "Why do you laugh?"; B: response to "What does it [laughter] do for you?")

I. Nurses

Meaning Units	Central Themes
A. 1. To enjoy life, myself, and my friends	1. To enjoy life, myself, and friends
2. Laughter/humor lightens stress and can make life (the hard times especially) more easy to adjust to and allows me to take changes in stride.	2. Lightens stress; makes life easier to adjust to; allows subjects to take changes in stride
3. Occasionally I laugh because it's easier than crying about a bad situation.	3. Easier than crying about a bad situation

B. 1 Laughter makes me feel more relaxed and eases stressful situation.

1. Feel more relaxed; eases a stressful situation

2. I like myself better when I laugh. I'm easier to live with and it makes me feel good.

2. Like self better when laugh; easier to live with; makes subject feel good.

II. Educators

A. 1. I laugh because it is a reflection of my feelings. I enjoy life and laugh readily as a result.

1. Reflection of feelings; enjoy life

2. I laugh at jokes, at circumstances, frequently (very) at myself.

2. Laugh at jokes, circumstances, self

B. 1. Laughter makes me feel "up" and relaxed and comfortable.

1. Feel "up"; feel relaxed; feel comfortable

2. I find it is contagious and I laugh with others and others laugh with me.

2. It is contagious.

To Potential Research Participants:

Your assistance in responding to the attached set of questions on laughter will further our research project into the phenomena of humor. To participate, please fill in the demographic data at the top of the page and then write out as fully a reply as needed to each question to describe your experience. If, however, for any reason you feel disinclined to participate, please return the questionnaire blank with or without an explanation.

Laughter and Humor

Age_____   Race_____

Sex_____   Education_____

Marital Status_____   Occupation_____

Why do you laugh?

What does it (laughter) do for you?

Thank you for your time and assistance.

Appendix C

One Counselor's Intervention in the Aftermath of a Middle School Student's Suicide: A Case Study

Jo Ann C. Alexander and Robert L. Harman

Reprinted from Alexander, J. A. C. & Harman, R. L. One counselor's intervention in the aftermath of a middle school student's suicide: A case study. *Journal of Counseling and Development*, 1988, *66*, 283–285. © ACA. Reprinted with permission. No further reproduction authorized without written permission of the American Counseling Association.

The authors discuss the application of Gestalt theory as a means of dealing with the surviving classmates of a student who committed suicide.

Four young people have died of suicide within the last month in our county. The second of these was Jason, a 13-year-old student in the middle school in which I am a counselor. The first death occurred on Valentine's Day, and Jason's followed by 3 weeks. The subsequent two deaths occurred in other parts of the county within a week of Jason's death. These events seem like a poignant validation of the "Werther" effect (Phillips, 1985)—the tendency of humans to imitate.

It is important for school counselors to have skills not only in programming for suicide but also for intervening in the aftermath of suicide. Existing literature, however, offers little to prepare counselors—particularly those in school settings—for this role. Researchers (Calhoun, Selby, & Faulstich, 1982; Calhoun, Selby, & Selby, 1982) have reported on the aftereffects of suicide, but few actually (Hill, 1984; Zinner, 1987) have discussed the ways in which a counselor might intervene.

In short, I had little from the professional literature to inform me when I learned of Jason's death. My task, as I identified it, was to help our young people grieve over Jason, to assist them in the process of letting go of him, and to minimize the likelihood of copycat suicides. I did not know what to expect in terms of their response to the news. I was coordinator of guidance in the school, in which we had three counselors, one per grade level, and approximately 1,000 students. I was assigned to the sixth grade, in which Jason had been a student.

Fortunately, I had been a Gestalt therapy trainee for 2 years. Also, I had some specific training in working with suicidally depressed adolescents and had done considerable reading in this area. It was with this preparation that I began my interventions. The approach described below should not be used by counselors without the support of comparable theory, knowledge, and skill.

After I decided that the most effective use of my time would be to work primarily with those 150 students with whom Jason had daily contact, I met with the faculty to prepare a consistent and appropriate school-wide response. We agreed not to eulogize Jason but to focus in public on our feelings of grief, shock, loss, fear, and even anger. We would not glorify his act, nor would we ignore that which we would miss about him.

In my work with students, I relied heavily on my knowledge of the theory and practice of Gestalt therapy. My task was to enhance students' awareness of their thoughts, feelings, and sensations about the death of a classmate and also to help them learn to express themselves in ways that might be more nourishing to themselves and to others at this time of trauma. With awareness, students might have more choice about how to respond both to Jason's death and to their feelings of isolation, hopelessness, and despair. My intent was to involve each student in his or her present experience in as many ways as possible; I began by visiting six of Jason's seven daily classrooms. Access to these classrooms was not difficult: Six teachers were delighted and relieved to accept my offer to work with their students; only two remained in the classroom to participate in my one class period intervention. One teacher chose to work with his students himself.

Class-sized Groups, Individuals, and Groups of Two

Jason chose to die with his goodbyes left unsaid. His act was abrupt and blunt. So as not to deflect from the quality of his act, I entered each classroom and announced, "I'm here today to help you say goodbye to Jason Davis. Jason is dead. He committed suicide. . . . He won't be back. . . . Where did Jason sit?"

Most suicides constitute an unfinished gestalt. In these cases, goodbyes are left unsaid, and the question of why a life was taken is left unanswered. Jason's was no exception. The purpose of my work, then, was to encourage students to say goodbye to Jason as a preface to letting go, to experience the collective and individual responses to his death in the here and now, to open avenues for intimate relating, and to explore constructive ways of coping with the situation.

Students acknowledged sadness and anger. My responses were intended to legitimize their feelings of betrayal and resentment. Those who had been the targets for some of his obscure signals were given the opportunity to cry and to speak to Jason's empty seat to tell him of their anger, resentment, betrayal, guilt, grief, confusion, sadness, and emptiness. Also, they were able to tell him what they would have done for him if they had known he was troubled. Others, as well, were given the opportunity to address Jason's empty seat, telling him what they would like him to know. Each was encouraged to end his or her statement with ". . . and goodbye, Jason."

For some, this experience seemed too threatening or overwhelming, so yet another mode of expression was offered: the nonverbal, subvocal goodbye. Students were invited to look at Jason's seat and imagine saying goodbye to him and to imagine telling him what they would like him to know. If time permitted, some classroom groups were given the

129

opportunity to write their goodbyes to Jason. The exercise was varied in the art class to allow for another avenue of expression, that of artistic representation of feelings.

Whenever a student exhibited strong emotion, I invited classmates to respond directly to that student. The students were exceptionally kind, caring, and supportive in their relating with each other. Many pleaded with their peers, "Please don't leave me. I'll help you." Others said, "I'm afraid I'll kill myself."

As the day progressed, I noticed that some students had been present in previous classes, so I invited them to remain in the classroom or gave them an opportunity to go to the library instead. Only one child, Jason's closest school friend and classmate in all of his seven classes, elected to go to the library. He did, however, choose to participate in five classroom sessions and requested two additional sessions, one of which is described in the section below.

At the conclusion of these classroom sessions, I offered the opportunity for additional counseling. As a result, several students sought individual or dyadic (students in pairs) sessions. Because of an expressed continuing need, I formed a small group that met once and another small group that met weekly for the remaining 3 months of the school year. The individual and dyadic sessions, as well as the small-group sessions, were similar to the work done in the classrooms but with more intense, focused attention.

Small-Group Sessions

Initial Group

The group sessions proved to be by far the most intense of the counseling sessions that I conducted. I employed with these students a projective technique adapted from that suggested by Oaklander (1978). Students were asked to think for 1 minute about Jason and his death. When time was called, they each were provided with a huge sheet of paper and some crayons, and they were asked to express their feelings on paper in colors, lines, shapes, and symbols. I paid attention to how each student approached and continued the task as well as to the picture itself. This proved valuable in helping students to reown previously disowned parts of themselves and to identify some who currently might be considered at risk.

The drawing of one student, Jason's closest school friend, seemed very simple and resembled the letters "JOI." In speaking as though he were each part of his drawing, he described his own feelings of emptiness, loneliness, and confusion, as well as his own suicidal fears:

This is the part of my brain that says, "Do it."

This is the part of me that says, "Don't do it."

I'm a hook with a sharp end. I can hurt you.

I'm going round and round.

I'm left hanging. I'm empty inside.

I'm straight and bright and happy when I don't think about Jason.

A portion of our subsequent time was spent on his belief that he must keep himself busy so that he would not think about Jason. His fear was that if he thought about Jason, he might hurt himself. Consequently this child was expending a tremendous amount of energy in his attempts not to acknowledge his feelings and was experiencing a great deal of anxiety. In the group setting, he was able to express his feelings in a safe environment and to receive caring and support from group members.

Subsequent Group

As a result of several students' expressed continuing need, I formed a long-term group and held weekly sessions for the remaining 3 months of the school year. This group was composed of six girls, three of whom had been in the initial group and wanted to continue. Because some closure had been reached for the other two members of the initial group, I formed another group, which was to meet for 12 weeks. During the first session, Lynne was observed tearing pieces from her notebook as she spoke about her sadness and confusion. In the Gestalt mode of staying with "what is," Lynne was invited to continue to tear her notebook and to see where that might bring her. When asked to give her hand a voice, she said, "I'm tearing up my notebook in little pieces." When asked if there was anyone in her life she would like to tear up, she replied, "Yes. Jason and me."

I directed her to "tear Jason up, tell him how you feel about his leaving you." After she completed her response to Jason, I invited her to become the pieces and give them a voice, at which time she described being "all torn up, broken, nothing but a pile of pieces . . . I should have done something to stop him. I knew. It's my fault. I hate myself."

Lynne seemed to be making progress on undoing her process of retroflecting (turning back onto herself) her anger and destruction when Anna tearfully interrupted, "This is the second time this has happened to me. My brother committed suicide." At this point, the focus of the group's attention turned to Anna.

During Anna's intensive work on her brother's suicide, many of her comments suggested that she believed her peers were laughing at her, thinking she was dumb. So that she could become aware of what was really out there, I invited her to look at each person and to tell me what she observed. She reported seeing each person looking at her and not laughing, but she was still imagining she was dumb.

I told her to "look at each person." At this point, she perceived much genuine warmth from the group members. In addition to the support being given to Anna, each student was now voluntarily holding hands with one or two other group members. The group ended with each girl looking directly at one or more of the other group members and clearly stating what she needed from that other person. Many said, "Don't be my friend and leave me like Jason did."

Many of our subsequent sessions proved to be as intense. During our third session, four of the six members revealed that they had attempted suicide. Two reported at least two prior attempts. The remaining two reported having seriously considered suicide.

Conclusion

I do not, unfortunately, know the actual impact of my work with these students. I do know, however, that I was deeply moved by their capacity for grieving and for caring for each other. Through our work together, I developed a great deal of caring for these young people, whose behavior had previously not drawn me to them.

Several themes emerged from my encounters with Jason's classmates. They were experiencing the various grief responses and a pronounced fear that others would follow Jason's example. Not only were they greatly afraid of being faced with the loss of yet another friend, but many also feared their own suicidal potential. The incidence of previous suicide attempts was alarming. The death of a friend highlighted several other issues as well: poor self-concept, excessive self-demands, fear of loss, grief over previous losses, self-blame, and self-recrimination.

Although this is not a study of the responses of teachers, administrators, and counselors in the school, my observation is that they feel unprepared to deal effectively with such a tragedy. Perhaps, consequently they are prone to avoid the issue. In this case, they seemed shocked and almost paralyzed. Most wanted someone else to handle the situation.

Generally, the students who characteristically exhibited such problem behaviors as skipping school and disrupting class were the most verbal participants. These students seemed "stirred up" by Jason's suicide. They were the risk takers again, but this time in a positive and healing way. They were the catalysts who brought their classmates together in more intimate, supportive, and caring ways.

Appendix C

References

Calhoun, L., Selby, J., & Faulstich, M. (1982). The aftermath of childhood suicide: Influences on the perception of the parent. *Journal of Community Psychology, 10,* 250–254.

Calhoun, L., Selby, J., & Selby, L. (1982). The psychological aftermath of suicide: An analysis of current evidence. *Clinical Psychology Review, 2,* 409–420.

Hill, W. (1984). Intervention and postvention in schools. In H. Seidak, A. Ford, & N. Rushforth (Eds.), *Suicide in young* (pp. 407–415). Littleton, MA: John Wright.

Oaklander, V. (1978). *Windows to our children.* Moab, UT: Real People Press.

Phillips, D. (1985). The Werther effect: Suicide and other forms of violence are contagious. *The Sciences, 25,* 32–39.

Zinner, E. (1987). Responding to suicide in schools: A case study in loss intervention and group survivorship. *Journal of Counseling and Development, 65,* 499–501.

Jo Ann C. Alexander is a former middle school counselor and currently is a counselor in private practice, Winter Park, Florida.

Robert L. Harman is director of the Counseling and Testing Center, University of Central Florida, and is in part-time practice in Orlando, Florida. Correspondence regarding this article should be sent to Jo Ann C. Alexander, Licensed Mental Health Counselor, 1155 Louisiana Avenue, Suite 212, Winter Park, FL 32789–2335.

Appendix D

Effect of Therapist's Self-Disclosure on Patients'
Impressions of Empathy, Competence, and Trust in
an Analogue of a Psychotherapeutic Interaction[1]

John M. Curtis[2]
Los Angeles, California

Reproduced with permission of author and publisher from: Curtis, J. M. Effect of therapist's self-disclosure on patients' impressions of empathy, competence, and trust in an analogue of a psychotherapeutic interaction. *Psychological Reports*, 1981, *48*, 127–136. © *Psychological Reports* 1981

Summary—The present study examined the relationship between a therapist's self-disclosure and the patients' impressions of the therapist's empathy competence, and trust. Written dialogues were constructed to manipulate three conditions of *high*, *low* and *no* disclosure by the therapist. 57 subjects were randomly selected and assigned to one of three treatment conditions, and the Barrett-Lennard Relationship Inventory and Sorenson Relationship Questionnaire were measures of perceived empathy, competence, and trust. Findings confirmed the initial predictions: the greater the use of therapist's self-disclosure, the lower the subjects' impressions and evaluations of the therapist's empathy, competence, and trust. The results raise doubt regarding the predictability of therapist's self-disclosure as a psychotherapeutic technique and suggest that, at least with respect to the type of self-disclosure used in this study, therapists who utilize self-dis-

1. This research was supported in part by a grant from the Beverly-Linden Mental Health Foundation.
2. Requests for reprints should be sent to John M. Curtis, Ph.D., 3443 Ocean View Boulevard, Glendale, CA 91208.

closing techniques may risk adversely affecting essential impressions on which a therapeutic alliance is established.

The use of therapist's self-disclosure has been part of the current controversy regarding the distinctions between counseling and psychotherapy. This controversy presumably began with Rogers (1951) who, in an attempt to de-emphasize the "medical model" influence borrowed from psychoanalysis, coined the term counseling to characterize more appropriately the psychotherapeutic endeavor.

Classical psychotherapeutic technique, which presumably originated with Freud (1912/1958), contraindicated the utilization of therapist's self-disclosure; instead, therapist's anonymity, i.e., the "blank screen" or "mirror" posture, and personal restraint (the "rule of abstinence") were recommended to help mitigate the contamination of the patients' transference reactions.

This preliminary caution was corroborated by many other psychoanalytic theorists and clinicians (Fenichel, 1941; Glover, 1955; Greenson, 1967; Langs, 1973, Menninger, 1958) by indicating that the therapist's expressed personal reactions tended to interfere with the analysis of the patients' transference discoveries and resolutions.

The emergence of non-psychiatric specialties as providers of psychotherapeutic service, a growing discontent with the genetic influence of psychoanalysis, as well as its limitations in terms of time, expense, and narrow range of patients to whom the treatment was applicable, and the escalating influence of behavioristic and humanistic-existential psychology, led to the development of new psychotherapeutic techniques.

In a marked departure from the traditional "blank screen," psychoanalytic posture, several theorists and researchers have identified what they deem to be essential therapeutic determinants: (a) Rogers (1957), the attitude of congruence; (b) Jourard (1964), who coined the term self-disclosure, the attitude of transparency; (c) Bugental (1965), the attitude of authenticity; (d) Kaisser (1965), the attitude of openness; and (e) Truax and Carkhuff (1967), the attitude of genuineness.

Several investigations (Davis & Skinner, 1974; Gary & Hammond, 1970;

Jourard & Resnick, 1970; Worthy, Gary, and Kahn, 1969) substantiate what Jourard (1971) designates as a "dyadic effect" of self-disclosure: that self-disclosures offered by the first party in a dyadic interaction elicit self-disclosures in the second party.

Other studies (Dies, 1973; Feigenbaum, 1977; Jourard & Friedman, 1970; Murphy & Strong, 1972; Vondracek & Vondracek, 1971) have shown that the use of therapist's self-disclosure favorably influences clients' perceptions necessary to the development of a strong therapeutic alliance; these findings, of course, are consistent with a humanistic-existential and/or behavioristic perspective.

In contradistinction to the aforementioned studies, however, are yet other investigations (Chaikin & Derlega, 1974a, 1974b; Derlega et al., 1976; Polansky, 1967; Truax et al., 1965; Vondracek, 1969; Weigal et al., 1972; Weigal & Warnath, 1968) which show that therapist's self-disclosure has adversely affected clients' impressions, i.e., perceptions and evaluations, regarding the therapist's "mental health" and professional comportment. These findings, of course, are consistent with the established recommendations of a psychodynamic orientation (cf. Freud, 1912/1958; Fenichel, 1941; Glover, 1955; Greenson, 1967; Langs, 1973; Menninger, 1958) and display noticeable equivocality in the available research.

Since the use of therapist's self-disclosure represents a salient distinction between psychodynamic and humanistic-existential paradigms of psychotherapy and because the inconsistencies found in previous research raise doubt as to the predictability of its [therapist's self-disclosure] technique, further research has become necessary.

A major objective of the present study was to help test the effectiveness of therapist's self-disclosure as a psychotherapeutic technique by achieving increased control over the independent variable. To accomplish this, following the research precedents of Rogers (1957) and Truax and Carkhuff (1967) on facilitative therapeutic conditions, the dependent variables of (a) empathy, (b) competence, and (c) trust were selected for investigation. These variables have been recognized to be essential inferences on which a therapeutic alliance is established, irrespective of the clinician's theoretical orientation.

It was hypothesized (p .05) that: (1) the therapist's *high* self-disclosure condition will yield the patients' lowest evaluations of empathy, competence, and trust; (2) the therapist's *low* self-disclosure condition will yield the patients' moderate evaluations of empathy, competence, and trust; and (3) the therapist's *no* self-disclosure condition will yield the patients' highest evaluations of empathy, competence, and trust.

Method

Subjects

Fifty-seven subjects currently receiving psychotherapeutic services, 29 of whom are male and 28 of whom are female, whose combined mean age was 32 years and ranged between 18 and 55 years, were randomly selected from a metropolitan mental health and family treatment agency. Subjects were asked to participate in a study designed to enhance services for patients to which they were randomly assigned to one of three conditions; these included three levels of therapist's self-disclosure (high, low, and none), the only experimental manipulation.

Design

A one-way analysis of variance design (see Kerlinger, 1973), employing three randomized groups of subjects, which corresponds to the three conditions of therapist's high, low, and no self-disclosure, was utilized.

Measures

Independent variable. Three written dialogues between a patient and therapist were constructed to serve as the stimulus conditions in which therapist's self-disclosure was manipulated across three separate conditions, i.e., the dialogues were specifically designed to vary systematically the corresponding levels of therapist's high, low, and no self-disclosure.

138

To accomplish this with optimal control over the independent variable, the patients' comments in all three dialogues were held constant, while only the therapist's remarks were varied systematically to reflect the change in the level of self-disclosure. Moreover, in order to assure this control the content of the therapist's remarks was held constant across all three dialogues; and only the form was altered by carefully changing the pronoun used in the therapist's responses.

Specifically the pronoun *I* was utilized in the therapist's responses in Dialogue I (e.g., I sometimes feel depressed too), the therapist's high self-disclosure condition. The pronoun *we* was utilized in the therapist's responses in Dialogue II (e.g., we all sometimes feel depressed), the therapist's low self-disclosure condition. The pronoun *it* was utilized in Dialogue III (e.g., it must have made you feel depressed), the therapist's no self-disclosure condition.

Further, to camouflage the subtle differences displayed in the three dialogues and to establish greater uniformity between the dialogues, three of the eight responses by the therapist were held constant by inserting three uniform, non-revealing, i.e., reflective, responses. Specifically, in Dialogue I, five of the therapist's eight responses included direct personal references; in Dialogue II, five of the therapist's eight responses included indirect personal references; and in Dialogue III, all eight of the therapist's responses included only reflective, non-revealing responses.

Dependent variables. The Sorenson Relationship Questionnaire[3] is a 24-item counseling-relationship questionnaire, using 7-point Likert-type scales for measuring the therapist's performance in the counseling situation. The instrument is divided into two parts, i.e., Part I measures empathy and Part II measures expertness. For the purposes of this study, both scales were utilized. Internal consistency reported was .85. The Barrett-Lennard Relationship Inventory (1962) is a 64-item counseling-relationship inventory, using 6-point Likert-type scales for measur-

---

3. A. G. Sorenson, Toward an instructional model for counseling. Occasional Report No. 6, Center for the Study of Instructional Program, University of California, Los Angeles, CA, 1967.

ing the four counselor/therapist traits of (a) regard, (b) empathy, (c) unconditionality, and (d) congruence. For the purposes of this study, only the empathy and unconditionality scales were utilized. Unconditionality, as defined by Barrett-Lennard (1962), was taken to represent a measure of trust. Split-half reliability reported as .82 (Barrett-Lennard, 1962).

Procedures

Subjects were randomly selected from a file of ongoing treatment cases and then randomly assigned to one of the three experimental conditions, i.e., the written dialogues. The subject's therapist was notified regarding the researcher's intention to enlist his patient in the study; written consent was obtained, and appointment times were scheduled.

The experimenter greeted each subject individually in the agency's waiting room just prior to his scheduled appointment time. The subjects were then asked if they would participate in a study being conducted by the agency to enhance services to patients. Upon agreeing to participate, the subjects were presented individually with the pre-selected dialogue and the two relationship questionnaires. Next, subjects were instructed to read the dialogue and obtain a general impression of the therapist's performance, and to evaluate this performance by completing the two distributed questionnaires. Subjects were also informed that, one week following the completion of their questionnaires, they would be invited to discuss any questions related to the research. This served to motivate the subjects toward conscientious participation. Subjects were debriefed as necessary.

Data Analysis

The data analysis included means and standard deviations of the dependent variables. Also, one-way analyses of variance (Myers, 1966) employing the Newman-Keuls multiple comparison procedure (Winer,

1962)—a *post hoc* test of mean differences—was utilized to analyze the sets of data.

Results

The means and standard deviations of the subjects' perceptions and evaluations of the therapist's relative levels of (a) two measures of empathy, (b) competence, and (c) trust are shown in Table 1.

Table 1. Means and Standard Deviations of Perceived Therapist's Empathy, Competence, and Trust under Three Levels of Therapist's Self-Disclosure

Measure of		Self-Disclosure		
Perceived		High	Low	N
Therapist				
Empathy*	*M*	29.68	30.05	37.73
	SD	6.23	7.27	9.37
Empathy II**	*M*	56.94	67.10	72.93
	SD	14.18	12.27	11.71
Competence†	*M*	58.57	67.42	75.52
	SD	13.20	10.10	8.67
Trust‡	*M*	30.57	31.36	38.63
	SD	7.76	8.59	8.45

* Barrett-Lennard's Empathy Scale of the Relationship Inventory (Barrett-Lennard, 1962).

** Sorenson's Empathy Scale (Relationship Questionnaire, Part I).[3]

† Sorenson's Expertness Scale (Relationship Questionnaire, Part II).[3]

‡ Barrett-Lennard's Unconditionality Scale (Relationship Inventory, Barrett-Lennard, 1962).

141

Appendix D

Table 2. Analyses of Variance of Therapist's Perceived
Empathy I[†] and II[‡]

Source	df	Empathy I		Empathy II	
		MS	F	MS	F
Between	2	392.75	6.21*	1216.64	15.53*
Within	54	63.23		78.34	
Total	56				

† Barrett-Lennard's Empathy Scale of the Relationship Inventory (Barrett-Lennard, 1962).

‡ Sorenson's Empathy Scale (Relationship Questionnaire, Part I).

* $p < .01$.

The one-way analysis of variance performed on Therapist's Perceived Empathy I (see Table 2) yielded an F of 6.21 (p .01). A multiple-comparison procedure of mean differences utilizing the Newman-Keuls *post hoc* test indicated that this effect was based on the differences between the following conditions: (a) a therapist's high self-disclosure and none (p.01); and (b) a therapist's low self-disclosure and none (p.05). The subject's highest evaluations of Empathy I, as predicted, occurred with no self-disclosure by a therapist.

The one-way analysis of variance performed on Therapist's Perceived Empathy II (see Table 2) yielded an F of 15.33 (p.01). A multiple-comparison procedure of mean differences utilizing the Newman-Keuls *post hoc* test indicated that this effect was based on the differences between the following conditions: (a) therapist's high self-disclosure and no self-disclosure (p.01); and (b) a therapist's low self-disclosure and no self-disclosure (p.01). The highest evaluations for Empathy II, as predicted, occurred with no therapist's self-disclosure.

Table 3. Analyses of Variance of Therapist's Perceived
Competence[†] and Trust[‡]

Source	df	Competence		Trust	
		MS	F	MS	F
Between	2	1365.11	7.388	374.36	5.17*
Within	54	184.92		72.32	
Total	56				

† Sorenson's Expertness Scale Relationship Questionnaire, Part II).

‡ Barrett-Lenard's Unconditionality Scale (Relationship Inventory, Barrett-Lennard, 1962).

* $p < .01$.

The one-way analysis of variance performed on Therapist's Perceived Competence (see Table 3) yielded an F of 7.38 (p.01). A multiple-comparison procedure of mean differences utilizing the Newman-Keuls *post hoc* test indicated that this effect was based on the differences between therapist's high self-disclosure and none (p.01). The subjects' highest evaluations for Competence, as predicted, occurred with no self-disclosure by a therapist.

The one-way analysis of variance performed on Therapist's Perceived Trust (see Table 3) yielded an F of 5.17 (p.01). A multiple comparison procedure of mean differences utilizing the Newman-Keuls post hoc test indicated that this effect was based on the differences between the following conditions: (a) high therapist's self-disclosure and no therapist's self-disclosure (p.05); and (b) low therapist's self-disclosure and no therapist's self-disclosure (p.05). The highest evaluations for Trust, as predicted, occurred with no therapist's self-disclosure.

Discussion

One of the specific findings of this study was that therapist's self-disclosure adversely affected the subjects' impressions of the therapist's empathy, competence, and trust: the *higher* the level of therapist's self-disclosure the *lower* the subjects' evaluations of the therapist's performance on these prescribed dimensions.

The results do not confirm data from a number of investigations (Davis & Skinner, 1974; Dies, 1973; Jourard & Jaffe, 1970; Nilsson, Strassberg, & Bannon, 1979; Vondracek & Vondracek, 1971) which have indicated that therapist's self-disclosure may have significant therapeutic value, both in terms of its beneficial effects on clients' impressions and on ultimate therapeutic outcome. Rather, the results run somewhat parallel to other studies (Chaikin & Derlega, 1974a, 1974b; Derlega et al., 1976; Polansky, 1967; Truax et al., 1965; Vondracek, 1969; Weigal et al., 1972; Weigal & Warnath, 1968) which have demonstrated unfavorable effects of disclosure by the therapist on clients' perceptions and attitudes.

Since clients' impressions influence the "dyadic effect" of self-disclosure, i.e., a reciprocal transaction process between counselor and client (see Jourard, 1971), it would be reasonable to expect that the clients' negative impressions of the therapist would unfavorably affect this exchange process. Further, in a number of studies which seem to corroborate a "dyadic effect" (Davis & Skinner, 1974; Gary & Hammond, 1970; Jourard & Resnick, 1970; Worthy, Gary, & Kahn, 1969) it is difficult if not impossible, to determine whether the effect of self-disclosure *per se*, or some other aspect of the interpersonal communication, such as similarity of attitude leading to liking, is partly responsible for some of the effects obtained. For instance, similarity of attitude (see Rubin, 1973) leads to liking which, along with the effect of a positive first impression (see Jones et al., 1968), affects subjects' inferences, i.e., attitudes and evaluations, on a variety of measures, e.g., likeability, professionality, rapport, "mental health," etc. Moreover, even Jourard and Jaffe (1970) recognized that research on self-disclosure is particularly sensitive to these "halo effects" which require increased control and precision. This was accomplished in this study by systematically varying the level of the therapist's self-disclosure in written dialogues.

Other recognized sources of potential invalidity to which research on self-disclosure is especially vulnerable are: (a) experimenter bias (see Rosenthal, 1966), a factor to which Jourard and Jaffe (1970) attribute many of the reported effects; (b) selection factors (see Campbell & Stanley, 1963) for how different samples, e.g., undergraduate college students or inpatient schizophrenics, affect predictions; and (c) demand characteristics (i.e., the role of subjects' expectations) such as a subject's preconceived attitudes which influence the outcome. These potentially contaminating factors were controlled in this study by utilizing written dialogues as the treatment conditions, and by selecting subjects currently receiving psychotherapeutic services—the group to which generalizability was targeted—as participants.

The present study was analogue in nature and cannot be regarded as an actual therapeutic encounter. The results might have proven somewhat different had subjects, having long-term affective involvement characteristics of a psychotherapeutic relationship, been evaluating their own therapist's performance. This potential limitation, however, needs to be assessed against the benefits obtained from increased experimental control.

Since the therapist's self-revealing responses are uniformly contraindicated by psychodynamically oriented treatments and since the results seem to be consistent with these established recommendations, these findings need to be interpreted in light of the specific type of therapist's self-disclosure used in the study, i.e., the change in the pronoun used in the therapist's remarks, which corresponds to high, low, and no disclosure by the therapist. Culbert (1970) identified several types of therapist's self-disclosure, e.g., anecdotal, experiential, feeling responses, etc., which, when incorporated into similar experimental conditions, may have yielded somewhat different results.

The fact that reciprocity has been recognized as a powerful determinant of dyadic interaction (see Homans, 1961; Jourard, 1971) and that a therapist's self-disclosure has been used to promote this kind of transaction cannot be regarded as conclusive evidence supporting its therapeutic utility. Aside from its appropriateness in "everyday" non-therapeutic, social interaction, the use of therapist's self-disclosure has been found in this and other research (Chaikin & Derlega 1974a, 1974b; Polansky, 1967; Vondracek, 1969) to yield unpredictable results.

Since the effect of first impressions (Jones et al., 1968) influences subjects' perceptions and evaluations and since the therapist's self-disclosure may promote patients' undesirable inferences of empathy, competence, and trust, this behavior may interfere with the establishment of rapport in a therapeutic relationship. At the very least, the effects of therapist's self-disclosure—if predictability may be regarded as a measure of utility—cannot yet be predictably determined across its many varieties and conditions.

Clearly, additional research is necessary, perhaps incorporating alternative varieties of therapist's self-disclosure—anecdotal, experiential, feeling responses, etc.—and contrasting effects between men and women, among different diagnostic groups, as well as between short- and long-term therapy. Such research may be helpful in determining the effects of therapist's self-disclosure under these prescribed conditions.

References

Barrett-Lennard, G. T. Dimensions of therapist responses as causal factors in therapeutic change. *Psychological Monograph*, 1962, *76*, No. 4 (Whole No. 562).

Bugental, J. T. *The search for authenticity.* New York: Holt, Rinehart & Winston, 1965.

Campbell, J. T., & Stanley, J. C. *Experimental and quasi-experimental designs for research.* Chicago: Rand McNally, 1963.

Chaikin, A. L., & Derlega, V. J. Looking for the norm-breaker in self-disclosure. *Journal of Personality and Social Psychology*, 1974, 42, 117-129. (a)

Chaikin A. L., & Derlega, V. J. Variables affecting the appropriateness of self-disclosure. *Journal of Personality and Social Psychology*, 1974, *42*, 588–593. (b)

Culbert, S. A. The interpersonal process of self-disclosure: It takes two to see one. In R. T. Golembiewski & A. Blumberg (Eds.), *Sensitivity training and the laboratory approach*. Itasca, NY: Peacock, 1970, pp. 88–100.

Davis J. D., & Skinner, A. E. G. Reciprocity of self-disclosure in interviews: Modeling of social exchange. *Journal of Personality and Social Psychology*, 1974, *29*, 779–784.

Derlega, V. J., Lovell, R., & Chaikin, A. L. Effects of therapist self-disclosure and its perceived appropriateness. *Journal of Consulting and Clinical Psychology*, 1976, *44*, 886.

Dies, R. R. Group therapist self-disclosure: An evaluation by clients. *Journal of Counseling Psychology*, 1973, *20*, 344–348.

Feigenbaum, W. M. Reciprocity in self-disclosure within the psychological interview. *Psychological Reports*, 1977, *40*, 15–26.

Fenichel, O. *The psychoanalytic theory of neurosis*. New York: Norton, 1941.

Freud, S. Recommendations to physicians practicing psychoanalysis. In *Standard edition of the complete psychological works of Sigmund Freud*. London: Hogarth Press, 1958. Vol. 12. pp. 109–120. (Originally published, 1912)

Gary, A. L., & Hammond, R. Self-disclosure of alcoholics and drug addicts. *Psychotherapy: Theory, Research, and Practice*, 1970, *7*, 151–171.

Glover, E. *The technique of psychoanalysis*. New York: International Universities Press, 1955.

Greenson, R. R *The technique and practice of psychoanalysis*. New York: International Universities Press, 1967.

Homans, G. C. *Social behavior: Its elementary forms*. New York: Harcourt, Brace & World, 1961.

Jones, E. E., Rock, L., Shaver, G. K., & Ward, L. M. Pattern of performance and ability attribution: An unexpected primacy effect. *Journal of Personality and Social Psychology*, 1968, *10*, 317–340.

Jourard, S. M. *The transparent self*. Princeton, NJ: Van Nostrand, 1964.

Jourard, S. M. *The transparent self*. (Rev.) Princeton, NJ: Van Nostrand, 1971.

Jourard, S. M., & Friedman, R. Experimenter-subject "distance" and self-disclosure. *Journal of Personality and Social Psychology*, 1970, *15*, 278–282.

Jourard, S. M., & Jaffe, P. E. Influence of an interviewer's self-disclosure on the self-disclosure behavior of interviewees. *Journal of Counseling Psychology*, 1970, *17*, 252–257.

Jourard, S. M., & Resnick, J. L. Some effects of self-disclosure among college women. *Journal of Humanistic Psychology*, 1970, *10*, 84–93.

Kaiser, H. *Effective psychotherapy*. New York: Free Press, 1965.

Kerlinger, F. S. *Foundations of behavioral science research* (2nd ed.). New York: Holt, Rinehart & Winston, 1973.

Langs, R. *The technique of psychoanalytic psychotherapy*. Vol. 1. New York: Jason Aronson, 1973.

Menninger, K. A. *Theory of psychoanalytic technique*. New York: Basic Books, 1958.

Mowrer, O. H. *The new group therapy*. Princeton, NJ: Van Nostrand, 1964.

Murphy, K. C., & Strong, S. R. Some effects of similarity self-disclosure. *Journal of Counseling Psychology*, 1972, *19*, 121–124.

Myers, J. L. *Fundamentals of experimental design*. Boston: Allyn & Bacon, 1966.

Nilsson, D. E., Strassberg, D. S., & Bannon, J. Perceptions of counselor self-disclosure: An analogue study. *Journal of Counseling Psychology*, 1979, *26*, 399–404.

Polansky, N. A. On duplicity in the interview. *American Journal of Orthopsychiatry*, 1967, *37*, 568–580.

Rogers, C. R. *Client-centered therapy*. Boston: Houghton Mifflin, 1951.

Rogers, C. R. The necessary and sufficient conditions of therapeutic personality change. *Journal of Consulting Psychology*, 1957, *21*, 95-103.

Rosenthal, R. *Experimenter effects in behavioral research*. New York: Appleton-Century-Crofts, 1966.

Rubin, Z. *Liking and loving: An invitation to social psychology*. New York: Holt, Rinehart & Winston, 1973.

Truax, C. B., & Carkhuff, R. R. *Toward effective counseling and psychotherapy: Training and practice.* Chicago: Aldine, 1967.

Truax, C. B., Carkhuff, R. R., & Kodman, F. Relationships between therapist-offered conditions and patient change in group psychotherapy. *Journal of Clinical Psychology*, 1965, *21*, 327–329.

Vondracek, F. W. The study of self-disclosure in experimental interviews. *Journal of Psychology*, 1969, *72*, 55–59.

Vondracek, S. I., & Vondracek, F. W. Self-disclosure in pre-adolescents. *Merrill-Palmer Quarterly*, 1971, *17*, 51–58.

Weigal, R. C., Dinges, N., Dyer, R., & Straumfjord, A. A. Perceived self-disclosure, mental health, and who is liked in group treatment. *Journal of Counseling Psychology*, 1972, *19*, 47–52.

Weigal, R. C., & Warnath, C. F. The effect of group psychotherapy on reported self-disclosure. *International Journal of Group Psychotherapy*, 1968, *18*, 31–41.

Winer, B. J. *Statistical principles in experimental design.* New York: McGraw-Hill, 1962.

Worthy, M., Gary, A. L., & Kahn, G. M. Self-disclosure as an exchange process. *Journal of Personality and Social Psychology*, 1969, *12*, 59–63.

Appendix E

The Monocultural Graduate in the Multicultural Environment: A Challenge for Teacher Educators

Mary Lou Fuller, *University of North Dakota*

I always thought I would teach in the Midwest. I had my whole life planned and those plans certainly didn't include teaching in Texas (Kim). Kim imagined herself teaching in an elementary classroom similar to those of her childhood. She would teach white, middle class children in a Midwest town or a suburb. But Kim and others graduating from colleges of education will find the classrooms in which they teach strikingly different from those of their childhoods. In this article I report on a study exploring what happens when monocultural students live and teach in multicultural environments and draw implications from those experiences for teacher education faculty and curricula.

Kim will find the class she teaches much more diverse from any she experienced as a student. By the year 2000 between 33% (Commission on Minority Participation in Education, 1988) and 40% (Hodgkinson, 1989) of school children will be from ethnically/racially/culturally/economically diverse groups. Increased immigration, higher birth rates for minorities and a declining number of white non-Hispanic children are the causes of these demographic changes (Griffith, Frase, & Ralph, 1989). Currently, the majority of students are children of color in 23 of the nation's 25 largest cities (Gay, 1989; National Center for Educational Statistics, 1987).

Kim will also find the economic status of her students' families may vary considerably from that of her family. Haberman (1989) predicts that one third will live in poverty by the year 2000. Kim will find the impact of poverty exacerbated because this population includes a large proportion of children six and younger—the most developmentally sensitive years—(Children's Defense Fund, 1989), with even higher proportions for Hispanic and Black children (Moneni, 1985).

Teacher education faculty must recognize the new demographics and identify and respond to their educational implications. They cannot assess the effectiveness of their professional practices without considering the needs of contemporary classrooms and teachers. An important criterion of the effectiveness of teacher education faculty is the teaching proficiencies of their graduates, an effectiveness seen by examining the ways education graduates relate to the changing school environment and population.

Little evidence of change in teacher preparation or teachers' classroom strategies exists, despite marked demographic changes in classrooms, a situation Sleeter and Grant describe as *business as usual* (1993, p. 18). They review how teachers teach diverse populations, what they teach, and how they group students; they find educational changes incongruent with demographic changes. Schools generally do not meet the needs of children from diverse populations.

The population of the public schools is changing, but that of colleges of education is not. Teacher education programs generally serve students coming from largely middle class homes (Webb & Sherman, 1989) with middle class sensibilities. The numbers of white, female, and middle class preservice teachers are increasing (Fuller, 1992a; Zimpher & Ashburn, 1992; and Webb & Sherman, 1989), while the number of teachers from diverse populations is decreasing (Zapata, 1988). Thus the teaching population is becoming more monocultural (Hodgkinson, 1989; National Education Association, 1987), while the student population is becoming more multicultural.

These primarily white, female, middle class preservice teachers from rural or suburban environments (Fuller, 1992a) have little exposure to people different from themselves. They also experienced little contact with minorities in the colleges of education they attended as these institutions generally reflect the same demographics as the students' communities. Though one might expect that the preservice curriculum would attempt to remediate the students' multicultural deficiencies (Mills & Buckley, 1992), with few exceptions, this is not the case (Zeichner, 1993; Fuller, 1992a, 1992b; Gwaltney, 1990; Griffith, Frase, & Ralph, 1989; Kniker, 1989).

Preparing preservice teachers for their future classrooms grows more complex as the school population becomes more diverse. Changing demographics require changing teacher education strategies; education faculty must consider the demographics of their graduates' classrooms and inform themselves of their graduates' experiences in these new environments. Documenting what happens as monocultural preservice teachers (students with limited exposure to diverse populations) begin teaching in multicultural settings was a goal of this study.

I explored the experiences of monocultural elementary education graduates teaching in multicultural environments; I looked for the accommodations of graduates of a Midwestern elementary education teacher education program working and living in environments different from those they experienced as children. The university is in a primarily monocultural area, and its students are similar to preservice teachers nationally (Zeichner, 1993; Fuller, 1992a; Kniker, 1989).

In the study I view one of the major goals of multicultural education as : . . . *reform [of] the school and other educational institutions so that students from diverse racial, ethnic, and social class groups will experience educational equity* (Banks, 1994. p. 3). Based upon this definition, I distinguish those who are *multicultural*—well informed about and have had meaningful experiences with people of diversity—from those who are *monocultural*—those who have not had meaningful experiences with diverse populations. I affirm the philosophical position of social reconstructionism which . . . *prepares future citizens [e.g., as preservice teachers] to reconstruct society so that it better serves the interest of all groups of people. This approach is visionary.*

153

Although grounded very much in the everyday world of experience, it is not trapped in this world (Sleeter & Grant, 1993, p. 210).

Methods and Data Sources

Teachers in the study were recent elementary education graduates from an upper Midwestern university. All had taken one of two courses: Multicultural Education or Introduction to Indian Studies. The Multicultural Education course is anthropological in nature, provides information about given cultures, develops an understanding of the concepts necessary to work with children from other cultures, and is social reconstructionist in philosophy. Approximately 3/4 of the teachers interviewed in this study had taken Multicultural Education. They also studied multicultural concepts in the Introduction to Education course which all had taken.

I contacted all graduates within three years of the study's start and screened them to determine if they were teaching in culturally diverse communities. I used the following criteria to determine three geographic areas to visit in this study: the presence of diverse cultural groups; the presence of small towns, suburbs, and cities; and a number of graduates teaching in the area.

I used two data collection methods, the first a broad protocol including many open-ended questions to interview the teachers. I taped all these interviews as well as those with personnel directors and administrators in the districts employing the teachers. In the latter interviews I sought their impressions of monocultural teachers from the upper Midwest. The second data collection procedure was field observations of the teachers in action and the school community. I tape recorded and transcribed these observations. The classroom observations and interviews of school personnel were to verify the accuracy of the teachers' self-reports. I identified recurring themes and considered important those which a third or more of the teachers mentioned (Slotnick, 1982). I noted less frequently cited ideas and experiences providing particular insights and used participants' voices to clarify issues and provide examples.

Participants

I contacted and screened 354 elementary education graduates to determine their geographic distribution and professional responsibilities. Ninety-one percent (321) responded, all but 29 being involved in professional activities. Respondents lived in 29 states and four foreign countries.

I selected teachers in three states to interview and observe. I visited four communities in central Texas, a metropolitan area in Nevada, and a metropolitan area and two rural areas in Arizona. I wrote letters to all teachers and followed up with phone calls requesting permission to interview them and observe their classrooms. I arranged to interview administrators and personnel directors in the same districts.

I interviewed 28 teachers and observed 26 teaching in their classrooms (schedule conflicts prevented the observation of the other two). All of the participants grew up in monocultural communities and identified their families as middle class. All grew up in intact families; one reported her parents divorcing during her late teens. Three were male; 25, female. Five taught kindergarten; six, first grade; three, second grade; one, third grade; one, fourth; one, fourth/fifth combination; two, fifth grade; two, seventh; one, eighth; four, special education; and one was a substitute teacher. I also interviewed five personnel directors and two elementary school principals.

Themes

Analysis of the transcripts reveals six themes: reasons for relocating to their present teaching environment; satisfactions related to the relocation; dissatisfaction related to the relocation; recognition of personal growth; advice to preservice teachers; and feelings about their move.

Reasons for the Move

Forty-two percent of the participants originally preferred to stay in the upper Midwest but moved elsewhere *because that's where the jobs were.* Twenty-nine percent moved to the Southwest because of personal relationships and another 29% because they wanted new experiences. Although they moved to a part of the country more culturally diverse than their home environment, none reported cultural diversity as a consideration in moving to the Southwest.

The following comments reflect the experiences and feelings the participants shared concerning their moves to the Southwest. Jennie (all names are pseudonyms) had not even considered leaving the Midwest until she acknowledged the reality of the job market. *I didn't pick Texas, Texas picked me. I thought for sure I was going to get a job in Minneapolis—after all I had applied. Late in the summer I went to Chicago to apply and on their computer they had three screens full of people with my last name (Olson) and there were three other applicants with my exact name, Jennie Olson. The moment of truth! School had already started and I had resigned myself to subbing for the year when I got a job offer in Texas.*

For Nancy, the move to the Southwest was even more serendipitous. *I didn't intend to be here. I went to the Career Fair, standing in what I thought was a long line for positions in the Twin Cities* [Minneapolis and St. Paul] *and when I got to the table I found I was in the line for a large city district in the Southwest. After standing in line all that time I thought I might as well interview. As the summer wore on I became desperate—I hadn't heard from anyone. Then one day* [the large city district] *called my home and talked to my mother who volunteered that while she was sure I was professionally organized my bedroom at home was always a mess. Needless to say I was surprised when they called back and offered me a teaching position. After arriving it only took me a few weeks to know I was where I should be.*

Those who moved to the Southwest because of personal relationships were female; relationships varied from married to engaged and other configurations. Sandy's husband's position dictated where she would teach: *My husband graduated in Airport Administration and applied in a variety of cities. It was agreed that I would apply wherever he got a position. I'm surprised, but I really like it here.* Mary Jane, by

contrast, moved to pursue a relationship. *I did my student teaching here because this is where my boyfriend was living and I figured that this was the best way to get a teaching position near him. I'm glad that I did because I really like it here although the job turned out to be much more interesting than the relationship.*
Several moved to the Southwest for new experiences. Neither culture nor geographic location was important in deciding where to live. Karen wanted to be away from home and independent: *I wanted new experiences; I wanted to go anywhere. When I grew up in* [a small town] *I just wanted to get out of Dodge big time. I just wanted to know what the rest of the world was like and teaching not only gave me a way to leave, it gave me permission to leave. It was something my parents could understand. After all the money they spent on my education, they wanted to see me employed.*

Satisfactions

Moving was just the beginning; once there, participants began accommodating to both teaching and the new environment. The second theme concerns teachers' satisfaction as a result of moving to the Southwest. The participants reported more satisfactions, both in number and quality, than any other theme. Professional satisfactions included their individual schools (86%), the cultural diversity of schools and communities (75%), their school districts (64%), and the school personnel (64%). Satisfactions in their personal lives included cultural diversity (75%), more things to do (64%), weather (54%), nice people (50%), living conditions (46%), shopping (42%), beauty of the locale (39%)

Professional Satisfactions. The people the participants met after their moves provided important satisfaction. One participant said, *They just think we* [teachers from Midwest] *are so great and we just think they are so friendly, warm and nice.* Interviews with administrators and personnel directors confirmed that they saw the participants as great teachers.

Participants were generally enthusiastic about people they met: *I love the people, I like my peers, my principal—she is wonderful, and the parents have been wonderful too* (Candy). *People down here are very nice. If you need questions answered they are always willing to help* (Glenn).

Cultural diversity was important. Margaret commented: *I think that I've grown up a lot because of my move. I've gained a lot more experience than I would have if I had stayed in* [state] *just basically because I'm teaching a group of kids that are a lot different than I am. I am constantly learning things from them. I'm finding out all kinds of interesting things about different cultures. . . . Their different holidays, and their lifestyles, what is important to them and their families. It is interesting to see some of the things that the parents I'm talking to see as important compared to what my parents viewed as important.* Jayne observed: *My classroom is about half black and half white. Also I have a little girl who is from the Philippines and lives with her mother who is deaf and a little boy from Korea who has limited English proficiency. It is a wonderful class and I think the diversity is part of what makes it so great.*

Sandy noted that appreciating another culture gave her a greater appreciation of her own, and she learned to see herself in different ways: *When you consider the characteristics of someone else's culture you can't help but think about your own.*

Personal Satisfactions. Participants reported satisfaction with the activities available in their personal lives, *the more moderate temperatures, the kindness of people, the living conditions, shopping, and the beauty of the locale.* One of the underlying factors in this category was the number of options available: many things to do, different places to go, many things to buy, and a range of people to meet.

Their personal lives were enriched in a variety of ways. Kathy's experience was representative of many. Initially anxious about her move, she later expressed a high degree of satisfaction with her choice. *I was so nervous, I had never been so far from my family but I fell in love with it as soon as I got here. I walked in the front door of the school and Bobbi* [principal] *gave me a hug and I felt that I was home. The staff became my family and the teacher in the next room my best friend.* Living conditions, shopping, beauty of locale were also important: *Going to buy clothes and having more than one option. Going out for the evening and having lots of choices of things to do. Meeting more and different kinds of people. Always having more than one opportunity whether it be for fun, necessities, or jobs. I guess the key word is choice* (Mary Jane).

Dissatisfactions

The third theme dealt with dissatisfactions which appeared in fewer areas and which participants reported less frequently than satisfactions. They identified five common concerns: poverty (52%), gangs (50%), lonesomeness for their families (46%), child abuse (42%), and their students' family problems (39%). They mentioned three other dissatisfactions specific to given areas: participants' personal safety (42%); the theme that you *can't trust people the way you can back home* (36%) [reported almost exclusively in urban areas, typically by those recently moving to the area]; extensive and inappropriate standardized testing [only teachers in Texas] (36%).

Professional Dissatisfactions. The participants believed cultural diversity enriched their lives, but other forms of diversity—particularly in their professional lives—disturbed them. Prominent among these were economic diversity (poverty) and diverse family structures (e.g., single parent homes). While not mentioning Maslow's hierarchy, participants repeatedly discussed poverty and family structures in terms of their students' hierarchy of needs. They talked of unmet physical needs, such as hunger, safety needs focusing primarily on the unpredictability of some students' lives, and affiliation needs often resolved through gang membership.

The participants were unprepared for the poverty they saw. All but eight had children of poverty in their classrooms in numbers ranging from a few to most children. Many teachers, like Carol, initially had difficulty recognizing poverty. *I was surprised when a number of my second grade students came to school hungry. They came to school with a lot of problems, but not enough to eat. I just wasn't prepared for that and at first I didn't even recognize what was going on.* Amber had difficulty recognizing the educational implications of poverty. *Sometimes I felt annoyed when my great lessons were met with a lack of enthusiasm and then I realized that it wasn't a lack of enthusiasm but rather a lack of energy. I had to remind myself that it is hard for kids to concentrate on math when their basic needs haven't been met.*

While sensitive to their students' needs, participants lacked understanding of poverty and spoke impatiently about the parents who were victims of poverty. Their observations often suggested *blaming the victim* (Ryans, 1971); some expressed the belief that if *they* just tried harder, *they* could overcome economic adversities. They generally lacked understanding of the nature and causes of poverty, but they did not lack concern for their students in poverty. They often expressed that concern as a desire to help by being there. Amber observed: *I think that you make more of an impact on the lives of these children than on the lives of more affluent children.*

The teachers' childhood familial structures differed dramatically from those of their students. They recognized this and reported that their lack of knowledge of and experience with different family structures constrained their understanding of the students' families. *An area of diversity, at least for me, is the various family structures in my classroom. Very few of the children in my room live with their original parents. They live with single moms, one with a single father, stepparents, a couple live with their grandparents, and one divides her time between her mother and father's home. They don't look like families that I grew up with or know. I have to keep reminding myself that just because they don't look like my family doesn't make them any less of a family. But these families do seem to have a lot more dif-*

ficulties (Kathy). The teachers had particular difficulty separating complications in students' lives caused by limited income rather than by family style. They often saw single-parent homes as the cause of a child's problems when in fact they were describing difficulties created by a lack of resources.

Gangs were a problem to varying degrees in all but one district and represented another professional concern. The participants were uniformly unschooled on this topic. Nancy was still trying to understand when she observed: *It affected me right away. During the new teacher meetings I was told that I couldn't wear a bandanna to school because gangs wear bandannas. My students can't wear anything with LA Raiders emblems on them because they are also gang symbols . . . . and my students threaten one another with their big brothers who have gang affiliations. My students are only in second grade and we aren't an inner city school.* Those who had inservice training on this topic were much more sophisticated in their understanding of gangs. Vicky and a few others considered issues such as why gangs are attractive to children. She observed: *I think that the attraction of the gangs is the element of belonging. I don't think they* [gang members] *belong to a family or have any real friends. I don't think that they see themselves as fitting in anywhere so consequently gangs are very attractive to them.*

Personal Dissatisfactions. Each personal item reflects an adjustment to a new environment: lonesome for their families, concerned about their personal safety, and feeling that they *Can't trust people the way you can back home.* Participants used their home environment as reference points in making sense of their new environment. While she talked of trust, Jill's tone was one of incredulity as she repeated the same story three times: *There are a lot of nice people here but then there are some that are not so nice. [Another issue is . . .] being trusted. You have to have a bank card to get a check cashed! Everything is checked and then checked again. Can you believe that I had to be fingerprinted twice to teach in this district?* Check-cashing cards are uncommon in her state.

Appendix E

Personal Growth

Another important theme was personal growth. With obvious pride and pleasure, participants talked often about growing and changing as a result of their moves; all but one saw the Southwest's cultural diversity as important to their growth. Sixty-eight percent reported greater independence than they would have had if they had not left home; 64% believed they were now much more open to new experiences. While cultural diversity was not an issue in moving to the Southwest, the participants most often cited it as contributing to personal growth. While one might expect those participants with the most children of color in their classrooms to be most affected by cultural diversity, all but two discussed cultural diversity at some length. *I feel more open and accepting of people. You lose your prejudice. . . . and you learn to be accepting of one another's differences* (Glenn). *I have really been dependent upon my parents even though I didn't live at home. I lived on campus but they were only a hop, skip, and a jump away. I've missed them a lot more than I thought I would. I've learned to be very independent and to make my own decisions* (Amanda). *I was ready to come. I like living somewhere larger so I think I've adjusted well. It was a growing experience. I think that I've grown more in the last eight months than I have at any other point in my life. I started a brand new job, I finished college, I moved into a new apartment, left my family and friends, and did it all in one week. I think that I adjusted well. I love it, I really do. There are times, though, that I miss my family* (Toni). Their growth was evident in their resentment of stereotyping of their students and their families. As they became better informed, they became increasingly resentful of cultural misconceptions. For example, many of the participants reported that folks back home would ask questions indicating they had preconceived, negative attitudes toward people of other cultures. Interestingly, the comments they reported resenting are not uncommon among monocultural preservice teachers. However, now that the participants were living in culturally diverse settings, they uniformly resented these comments and questions. Toni noted: *Some-*

one said to me, 'I suppose they steal and all that stuff.' My Euro-American parents view me as a babysitter but my Mexican families see me as someone special; they call me teacher as a form of respect. They are really a gentle culture and not like what you see on TV where they are portrayed as thieves.

Twenty-six of the 28 participants attributed growth primarily to cultural diversity. Their interviews and classroom behaviors support their assessments. One participant appeared genuinely unaffected by the cultural diversity in her environment; another did not understand its significance.

Jackie, who was unaffected, had created in her very multicultural community an environment closely duplicating her upper Midwest roots. After many job interviews, she accepted a position in an upper middle class, white school and moved into an exclusively white, upper middle class apartment complex. She did not travel outside her part of the community or experience the cultural diversity her city offered. Jackie also expressed discomfort with the culturally diverse schools, people, and neighborhoods in her city. Her classroom mirrored perfectly the upper Midwestern school where she student taught except for the two non-English speaking students who were bused from another part of the city and were invisible in her classroom. She vaguely smiled in their general direction, but she did not acknowledge them in any other way, did not make eye contact with them, did not call them by name or speak directly to them during the several hours she was observed teaching. Nor, predictably, did the other students. Jackie did not provide these students—or their classmates—with the opportunity to . . . *draw people into a public place* (Brown, 1992, p. 8); she did not allow inclusion (Brown, 1992) for the two Hispanic students in her class. Jackie established a monocultural niche in her multicultural community and so missed the growth other participants experienced. Jackie showed that for change to occur, both the presence of cultural diversity and the desire to experience that diversity are required. Just being there is not enough.

Terri also missed the significance of life in a multicultural environment. Only two of the students in Terri's classroom spoke English as their first language and many others were first generation Americans. Her classroom was pleasant, but nothing in it or in Terri's teaching strategies reflected the cultural background of her students. Asked about the cultural characteristics of the students, Terri insisted there were no differences between her present students and those in her (upper Midwest) student teaching experience. She did not see the cultural differences. She likes her students; unfortunately, she is not culturally sensitive.

Advice for Preservice Teachers

The last two themes emerged in response to specific questions. First, I asked the participants what advice they had for a preservice teacher who might want to teach in a classroom next to theirs. Though they offered a variety of suggestions, they mentioned only six frequently enough to be noted: Learn Spanish (93%); take field experiences in multicultural environments (82%); take the Multicultural Education course (78%); take the Classroom Management course (57%); learn about families (50%); and be open minded and flexible (46%). Concerning learning Spanish, even those participants who had no Spanish speaking students believed it important. John said: *I would tell everyone to take Spanish. Spanish has become the second language in this country and it will help you personally and certainly help you professionally.* Jayne noted the importance of multicultural education on finding herself in a small town in Texas: *I remember thinking to myself, OK, now what did I learn in Multicultural Education? What am I supposed to do? The first thing I remembered was this is their culture and I must respect it. I need to respect them and their culture and not try to use my culture as a yardstick for other people's behaviors. Then I remembered [the professor] saying over and over that it was my responsibility to be well informed about my students and their culture. I still had some big adjustments to make but I felt secure that I could do it.*

Of the six items, five had preservice curricular implications while the other item concerned personal growth. Interestingly, while everyone discussed concerns about their students' families, only half saw this as an area in which students could prepare themselves. Similarly, the majority of the interviewees expressed concern about poverty and gangs and yet no one mentioned either subject for study. Evidently, the participants did not see these topics as subjects for inclusion in a teacher preparation curriculum.

Feelings about the Decision to Move

All twenty-eight said they were pleased with their decisions to move to the Southwest. Twenty-one (75%) agreed without any qualifications; six (21%) said that they might move home at some point in time; one person observed that it took her about 18 months before she felt pleased with her life in the Southwest.

Several of the teachers have moved since arriving in the Southwest, and several others were planning to move. All of the moves had been or were anticipated to be within the Southwest. In general, those contemplating moves were single and were moving to enhance their social lives.

The Six Themes in Summary

From the six themes (reasons for moving, satisfactions, dissatisfactions, personal growth, advice to preservice teachers, and assessment of their moves to the Southwest) emerged a pattern indicating the participants viewed positively their moves to more diverse areas of the country. Although ethnic/racial/cultural diversity was not a factor in their decisions to move to a multicultural community, it became the single most important element of their experience. They generally functioned very well in their classrooms and grew personally and professionally. Their professional concerns included: lack of knowledge and experience with diverse populations, diverse family structures, poverty, gangs, and child abuse. Since these are all curricular issues, thoughtful teacher education faculty members can address them.

The participants' personal satisfactions included pride in their independence and growth, enjoyment of the friendliness of new colleagues and acquaintances, and the more relaxed work environment. The personal dissatisfactions (missing family, concern for safety among those living in metropolitan areas, and feelings of mistrust) are not topics faculty can effectively address.

Educational Implications

The participants believed that the multicultural environments enhanced their personal and professional lives. While they were appreciative of and sensitive to the cultural diversity in their classrooms, my observations suggest that their teaching strategies were generally not culturally informed. They had profited from their preservice multicultural courses but lacked the ability to select appropriate teaching strategies for the environments. Preservice programs must provide the necessary information and should include multicultural field experiences. Such field work will provide the frame of reference preservice teachers need. It is difficult to internalize specific strategies when one lacks experiential reference. The participants themselves identified the need for more preservice multicultural field experiences.

College of education programs must expand their views of diversity to include the effects of social class and families. First, the curriculum should include issues such as poverty, family study, and additional multicultural training (including field experiences as noted) to better prepare students to work in multicultural environments. This will provide preservice teachers with the broad view notion of diversity (Zimpher & Ashburn, 1992) necessary to prevent parochialism.

Second, preparing preservice teachers for their move from a monocultural and small city environment to a multicultural and metropolitan one will ease their transitions. These two changes, taken together, will allow new teachers to focus their professional concerns sooner on the children they teach.

Failure to implement these changes will be very costly to the educational system generally and to the education of children of color particularly. Elementary teachers increasingly represent mainstream society, and they are generally not well prepared to teach diverse student populations. To counter this problem, colleges of education must simultaneously continue to recruit preservice teachers of diversity while carefully preparing the primarily white, middle class teachers who are and will continue to be the teachers for most children of color. The education of white, middle class teachers must prepare them to understand and appreciate diversity, as well as identify and use the appropriate instructional strategies for their students. Only in these ways will schools be able to address the needs of the diverse populations they serve. Failure to do this, in Villegas' view (1991), will result in serious difficulties: *It seems clear from the research that unless teachers learn to integrate the cultural patterns of minority communities into their teaching, the failure of schools to educate children will continue* (p. 19).

Consequently, preservice teachers must be knowledgeable in the strategies that best meet the academic needs of their students and for colleges to better prepare preservice teachers for diversity. The following is a list of specific strategies effective teachers of minority students use (Irvine, 1992):

- Have appropriately high expectations for students;
- Employ many different instructional materials and strategies;
- Use interactive rather than didactic methods;
- Use the students' everyday experiences in an effort to link new concepts to prior knowledge;
- Help students become critical thinkers and problem solvers.

Zeichner (1993) suggests the following bearing on preparation of preservice teachers:

- Students are helped to develop a clearer sense of their own ethnic and cultural identities;
- Students are taught about the dynamics of prejudice and racism and about how to deal with them in the classroom;

167

- Students are taught about the dynamics of privileges and economic oppression and about school practices that contribute to the reproduction of societal inequalities;

- Students are taught various procedures by which they can gain information about the communities represented in their classrooms;

- Students complete community field experiences with adults and/or children of another culture;

- Students live and teach in a minority community (immersion);

- Instruction is embedded in a group setting that provides both intellectual challenge and social support.

Implementing these changes to the preservice curriculum should make graduates such as those interviewed in this study more attuned to the needs of and approaches to all their students. The result will be teachers able to both help their students and themselves. They may build on the insight Jennie expressed so well: *It would be so boring if all of my students were white and middle class. I'm so glad that I'm here. Moving made me look past myself at other people and other cultures.*

References

Banks, J. A. (1994). *Multi-ethnic education* (3rd ed.). Boston: Allyn Bacon.

Brown, C. E. (1992). Restructuring for a new America. In M. E. Dilworth (Ed.). *Diversity in teacher education: New expectations.* San Francisco: Jossey-Bass.

Children's Defense Fund. (1989). *A vision for America's future.* Washington, DC: Children's Defense Fund.

Commission on Minority Participation in Education and American Life. (1988). *One third of a nation.* Denver: American Council on Education and Education Commission of the States.

Fuller, M. L. (1992a). Monocultural teachers of multicultural students: A demographic clash. *Teacher Education, 4*(2).

Fuller, M. L. (1992b). Teacher education programs and increasing minority school population: An educational mismatch. In C. A. Grant (Ed), *Research directions for multicultural education: From margin to mainstream*. London: Falmer Press.

Gay, G. (1989). Ethnic minorities and educationally equality. In J. A. Banks & C. A. Banks (Eds.), *Multicultural education: Issues and perspectives* (pp. 167–188). Boston: Allyn Bacon.

Griffith, J. E., Frase, M. J., & Ralph, J. H. (1989). American education: The challenge of change. Washington, DC: *Population Bulletin, 44*(4), 16.

Gwaltney, C. (1990). Almanac: Facts about higher education in the nation, the states, and D.C. *The Chronicle of Higher Education*, 11–29.

Haberman, M. (1989). More minority teachers. *Phi Delta Kappan, 71(10)*, 771–76.

Hodgkinson, H. L. (1989). *The same client: The demographics of education and services delivery systems*. Washington, DC: The Institute of Education.

Irvine, J. J. (1992). Making teacher education culturally responsive. In M. E. Dilworth (Ed.), *Diversity in teacher education: New expectations*. San Francisco: Jossey-Bass.

Kniker, C. R. (1989). *Preliminary results of a survey of Holmes and non-Holmes group teacher education programs*. Midwest Holmes Group, Chicago.

Mills, J. R., & Buckley, C. W. (1992). Accommodating the minority teaching candidate: Non-black students in predominantly black colleges. In M. D. Dilworth (Ed.), *Diversity in teacher education: New expectations* (pp. 134–159). San Francisco: Jossey-Bass.

Moneni, J. A. (1985). *Demography of racial and ethnic minorities in the United States: An annotated bibliography with a review essay*. Westport, CT: Greenwood Press.

National Center for Educational Statistics. (1987). Washington, DC: U.S. Printing Office.

National Education Association. (1987). *Status of the American public school teachers, 1985-1986*. Washington, DC: National Education Association.

Ryans, W. (1971). *Blaming the victim.* New York: Vintage Books.

Sleeter, C. E., & Grant, C. A. (1993). *Making choices for multicultural education* (3rd ed). New York: Merrill.

Slotnick, H. B. (1982). A simple method for collecting, analyzing, and interpreting evaluative data. *Evaluation in the health professions, 5*(3), 245–58.

Villegas, A. M. (1991). Culturally responsive pedagogy for the 1990s and beyond. *Trends and issues paper No. 6.* Washington, DC: ERIC Clearinghouse on Teaching and Teacher Education.

Webb, R. B., & Sherman, R. R. (1989). *Schooling and society* (2nd ed). New York: Macmillan Publishing Company.

Zapata, J. (1988). Early identification and recruitment of Hispanic teacher candidates. *Journal of Teacher Education, 39,* 19–23.

Zeichner, K. M. (1993). *Educating teachers for cultural diversity.* East Lansing, MI: National Center for Research on Teacher Learning.

Zimpher, N. L., & Ashburn, E. A. (1992). Countering parochialism in teacher candidates. In Dilworth, M. D. (Ed.), *Diversity in teacher education: New expectations* (pp. 40–59). San Francisco: Jossey-Bass.

Appendix F

Teacher Reactions to Behavioral Consultation: An Analysis of Language and Involvement

Mary M. Rhoades and Thomas R. Kratochwill
University of Wisconsin—Madison

Reproduced with permission of the publisher from: Rhoades, M. M., & Kratochwill, T. R. Teacher reactions to behavioral consultation: An analysis of language and involvement. *School Psychology Quarterly*, 1992, 7(1), 47–59.

Abstract

Explored [are] two dimensions of behavioral consultation that can potentially influence teachers' reactions to the consultation process. Two independent variables (consultee involvement and consultant language) were completely crossed to create four videotape scenarios differing only with respect to the manipulated variables. Elementary school teachers (N = 60) were randomly assigned to view and rate one of the four scenarios on a measure of acceptability. Subjects reported high ratings for technical language when the psychologist took a directive role and did not involve the teacher in the problem-solving process. Results are discussed within the context of previous acceptability research and future research concerning consultee involvement in the consultation process.

To provide services effectively to the greatest number of students, school psychologists often focus their intervention efforts on the teacher

The authors wish to thank Drs. Maribeth Gettinger, Frank Baker, and Steve Elliott for their helpful feedback on the study and participating teachers in the various school districts.

through behavioral consultation (Bergan & Kratochwill, 1990). Research has indicated that behavioral techniques are often effective, but that they are sometimes misinterpreted and evaluated by teachers as less acceptable than other techniques (Elliott, 1988). The link between acceptability and effectiveness has been demonstrated in recent research, and practitioners are recognizing the need to explore consumer reactions when implementing interventions (Reimers, Wacker, & Koeppl, 1987). Consumer acceptability of psychological interventions is also important from a legal and ethical standpoint in establishing the social validation of a technique and in understanding individual needs.

Requests for reprints along with a self-addressed stamped envelope should be sent to Thomas R. Kratochwill, School Psychology Program, Department of Educational Psychology, 1025 West Johnson Street, University of Wisconsin-Madison, Madison, WI 53706.

Despite attempts to make behavioral strategies more acceptable to consultees and more efficient in managing classroom difficulties, consulting school psychologists still encounter teacher resistance as a major obstacle to effective service delivery (Witt & Elliott, 1985; Witt & Martens, 1983). Two psychologist-mediated variables may have a potential effect on teacher acceptability. These variables include the degree to which the teacher is *involved* in a collaborative problem-solving process with the consultant and the amount of *technical language* used by the consultant in communication with the teacher (Elliott, 1988).

Researchers investigating why the bias against behavioral techniques exists, or how it might be changed, have not presented conclusive results, but have suggested a number of variables that should be considered. Woolfolk, Woolfolk, and Wilson (1977) found that college students viewing identical videotapes of teaching strategies rated those strategies labeled "behavior modification" as less effective than videotapes labeled "humanistic." A second study examined whether the presentation of a rationale for efficacy or a softening of behavioral terms would influence ratings (Woolfolk & Woolfolk, 1979). Results indicated that videotaped behavioral teaching strategies presented with rationales and humanized terms were rated favorably, but only by an undergraduate group. The behavioral language of the technique did, to some degree, determine raters' preferences, but the actual bias against behavioral teaching strategies was not clearly defined.

Medway and Forman (1980) expanded the work of Woolfolk et al. (1977) to the area of consultation by presenting videotapes of mental health and behavioral consultation between a school psychologist and a teacher to actual teachers and school psychologists in the field. Results of this study indicated that although psychologists preferred the mental health technique, teachers rated the behavioral model as more effective.

A series of three experiments using written case descriptions of teaching methods evaluated by undergraduate students was completed by Kazdin and Cole (1981). The potency of the labeling effect was questioned as a causal variable in the negative evaluation of behavior modification techniques. It was shown that the negative evaluation received by the behavior modification condition, as compared to humanistic and neutral conditions, was due primarily to the content of the method, rather than the label applied. It was suggested that past research into the negative evaluations of behavioral techniques may have misplaced emphasis on the importance of the label alone and that other variables be examined.

The impact of labeling bias was explored by Witt, Moe, Gutkin, and Andrews (1984). Case descriptions of classroom interventions evaluated by teachers indicated that a pragmatic description was rated as more acceptable than behavioral or humanistic alternatives. The behavioral description, emphasizing that "staying in at recess involved the contingent application of punishment for the explicit purpose of controlling the child's inappropriate behavior" (Witt et al., 1984, p. 364), was evaluated as least acceptable, especially when rated by more experienced teachers. It appears that the acceptability of equivalent interventions may be at least partially determined by the language used in the approach.

A review of research in this area indicates a potential bias against behavioral techniques, but only two studies used actual subjects from the field (Medway & Forman, 1980; Witt et al., 1984), and most focused on teaching techniques as opposed to school-based consultation (Kazdin & Cole, 1981; Woolfolk & Woolfolk, 1979; Woolfolk et al., 1977). Studies completed by Kazdin and Cole (1981) and by Witt et al. (1984) suggest a need to explore separately the issues of content and technical language in evaluating the bias against behavioral techniques. A stronger recommendation for such analyses could be made if future work in the field with videotaped scenarios, as opposed to written case descriptions, supported such hypotheses.

Active involvement of the consultee in the consultation process may be perceived as important because the teacher is in a unique position to provide perspectives on the utility of the intervention and possibly, ownership of intervention plans by the teacher will facilitate intervention integrity (Gutkin & Curtis, 1990; Witt, 1990). Reinking, Livesay, and Kohl (1978) reported that consultee implementation of programs developed during consultation are related directly to consultee involvement. It is often concluded that consultees prefer collaborative, rather than expert, consultation styles (Babcock & Pryzwansky, 1983; Fine, Grantham, & Wright, 1979; Wenger, 1979) and that solutions developed through collaborative consultation are more acceptable than those generated alone or by others (Fairchild, 1976; Reinking et al., 1978). But results are not clear cut. For example, Wenger (1979) examined teacher responses to a consultant's attempt to facilitate either a collaborative or an expert consultation relationship. The collaborative consultant involved the teacher in the process of determining the child's needs and in developing strategies and techniques for classroom interventions. Although the expert consultant condition included teacher involvement (e.g., input, perceptions, hypotheses), the consultant developed the recommendation and gave them to the teacher. The teachers exposed to the collaborative approach were more satisfied (as measured by ratings), but there were no significant findings on the recommendation implementations. The design was also quasi-experimental.

In the case of the Babcock and Pryzwansky (1983) study, three groups of education professionals (elementary school principals, special education teachers, and second-grade teachers) rated their preference for four consultation models as offered by school psychologists at five stages of consultation. The professionals rated the collaboration model the highest on a rating scale. But, finding a stage by model interaction led these authors to conclude that consultation preference should not be considered a unidimensional concept.

More recently Wiese and Conoley (1989) reported that perceptions of problem-solving style and exposure to a collaborative or expert consultation model did not significantly affect consultee self-reported problem-solving behaviors or expectations regarding a problem solution. Ratings of acceptability were used following undergraduate viewing of one of two videotapes depicting the two consultation conditions.

Involvement has also been examined from the perspective of a relational communication analysis. Erchul (1987) assessed consultants and consultees on two measures of relational control, domineeringness, and dominance. His results indicated that consultants controlled the dyadic relationship across all stages of behavioral consultation. Moreover, consultants having high dominance scores tended to be judged as more effective by consultees. Results of the study by Erchul and Chewning (1990) also indicated that in dyads where the consultant is dominant and the consultee is submissive, consultation outcomes are considered more positive by both parties. These authors suggest that behavioral consultation consists of a more cooperative than collaborative relationship.

To provide further information on the complexity of the consultant-consultee relationship, we were interested in examining two variables that have been studied in various contexts and dimensions in previous research, but not in behavioral consultation. Specifically, we assessed consultee involvement and consultant language in a completely randomized design for their effects on a measure of acceptability.

Method

Subjects

Participants included 60 white regular education teachers from 15 public elementary schools (K–6) in a Midwestern metropolitan area of approximately 170,000 people and three smaller elementary schools in three rural communities.[1] Each school enrolled 300–500 students, employed 16–25 full-time teachers, and included at least one special education program for students with exceptional learning needs. Only teachers with access to school psychological services at their schools were included. Pupil services staff working on-site at each school included the school psychologist, social worker, and guidance counselor. Subjects included 53 females and 7 males and half had at least 16 years of teaching experience. None of the teachers had training in behavioral consultation. Most participants were either relatively new or long-time veterans at their present school. Teachers who had been teaching in their present school placement for 1–3 years made up 27% of the subject sample while 37% of the subjects had been teaching at their present school for 16 years or more. Participants were adequately distributed across each grade placement as follows: K = 3, 1 = 13, 2 = 9, 3 = 9, 4 = 15, 5 = 6, and 6 = 5.

---

1. Readers might inquire about the number of schools that we had to contact to obtain the 60 subjects. We speculated that there were two reasons for needing this extensive sampling procedure. First, schools required us to submit information to teachers through building principals, who in turn, asked teachers to volunteer. Thus, we had no direct presentation to teachers to stimulate interest. Second, the schools are in an area where requests for subjects from a large research university were frequent and intense.

Procedures

After district approval to conduct research in the schools had been obtained, elementary principals were contacted by letter and phone to request school participation. When participation was secured, teachers were contacted by letter and asked to volunteer approximately 30 minutes of their time to participate in the viewing and rating of a videotaped consultation intervention. Consenting teachers were then scheduled for a viewing at their convenience.

Subjects were assigned randomly to one of four conditions: technical language with teacher involvement, technical language without teacher involvement, nontechnical language with teacher involvement, or nontechnical language without teacher involvement. Participants viewed one of the four taped scenarios of behavioral consultation between a female school psychologist and a female classroom teacher.

Prior to viewing the tape, subjects were given a brief introduction to consultation and their role in assessing the consultation scenarios. Teachers were told that the videotape they were about to view depicted a process for identifying and analyzing a problem and devising a plan for dealing with the child's problem in the classroom. They were also informed that other teachers would view a tape with a different style of consultation with the psychologist. Teachers were told that they should rate the acceptability of the consultation procedure and not the specific behavioral plan for the child portrayed in the tape. Copies of this introduction were given to each subject and were also read aloud by the experimenter. After viewing the consultation scenario, teachers were asked to complete the IRP–15 (Martens, Witt, Elliott, & Darveaux, 1985) which included a Likert rating of 15 items designed to assess acceptability of the consultation procedure. Teachers then completed six additional questions related to their teaching experience. Specifically, information was obtained on gender, years of teaching experience, years at present school, grade taught, use of the school psychologist, exposure to exceptional children, and satisfaction with school psychological services.

Videotapes.[2] Participants in each condition viewed one 12-minute video-taped scenario of consultation interactions between a teacher and a school psychologist. Scenarios depicted condensed versions of three of the four consultation stages outlined by Bergan and Kratochwill (1990), including Problem Identification, Problem Analysis, and Treatment Implementation. The consultation process was condensed to 12 minutes to sample the consultation process and to help ensure teacher participation. The fourth stage of consultation, Treatment Evaluation, was *not* included as the presentation of an intervention outcome would bias viewers who may change their rating based on specific intervention outcome data.

During the first consultation session, the consultant (psychologist) and consultee (teacher) identified the problem and set procedures for preassessment. The target problem in the videotape depicted an elementary school boy who demonstrated disruptive classroom behavior, including talking, arguing, disrupting others, and not completing work. In the second session, problem analysis occurred and a treatment was devised that included procedures for implementation in the classroom and an agenda for monitoring change. The intervention developed for the problem involved a note home–program with both parental and teacher reinforcement.

The target problem and intervention were identical across the four conditions. To assure that videotaped scenarios differed only with respect to the manipulated independent variables, participants followed scripts in which the basic content of the sessions described above and verbalizations between consultant and consultee were identical in all four conditions. A basic script was followed for each condition which varied only in the content of the condition. The participants role-playing consultant and consultee had experience in their respective roles. Both had been trained in behavioral consultation and both had been prac-

---

2. Copies of the videotapes and transcripts are available from the authors for the cost of reproduction and photocopy.

ticing school psychologists. Participants, physical setting, seating, filming, and problem content also were identical in all four scenarios. To assure identical filming, a preset sequence of close-up, individual, and wide angle shots was followed when filming each scenario. Cues within the scripts assured that nonverbal interactions (e.g., gestures) were also identical.

Prior to being viewed by teachers, videotaped consultation scenarios were also screened by 10 school psychology graduate students and 10 practicing school psychologists who served as blind raters. These preservice and inservice school psychologists had specific training in behavior modification and behavioral consultation and their ratings were used to assess whether the intended variables (i.e., consultant language and consultee involvement) were adequately represented and could be discriminated. The graduate students completed a 28-item rating scale assessing the content of each videotape to determine adequate portrayal of technical language, nontechnical language, teacher involvement, and teacher noninvolvement within each condition. The measure included an equal number of items from the consultee and consultant perspectives. The rater was asked to respond to each item on a Likert scale from 1 (extremely so) to 5 (not at all). There were 14 items representing involvement-noninvolvement (e.g., To what degree was the teacher actively involved in planning the intervention?) and 14 items representing technical/nontechnical language (e.g., To what degree did the psychologist rely on technical terminology to interpret the child's difficulties to the teacher?). Means computed on the pilot ratings of each videotaped scenario indicated that the independent variables (i.e., language and involvement) were present as intended. The school psychologists were asked to complete a 16-item checklist that required discrimination among the four conditions. For each scenario presented, the rater had to check the consultant language (technical or nontechnical) or teacher involvement (involved or noninvolved) condition represented. Results of this analysis indicated that the 10 psychologists discriminated the conditions with perfect accuracy.

179

Appendix F

Independent Variables

Teacher involvement versus teacher noninvolvement. Teachers' preferences toward high versus low degrees of teacher involvement were compared. Some subjects viewed a videotape in which the teacher was highly involved. Other subjects viewed tapes in which teacher involvement was minimal. Under the involvement condition, half the subjects viewed scenarios where the psychologist used technical terms. Other subjects viewed scenarios in which the psychologist used nontechnical language.

In the scenarios depicting teacher involvement, the consultant elicited and utilized the teacher's own ideas to devise a treatment plan and formulate an intervention. In the case of the involvement variable, the consultant, for example, stated during the problem identification interview, "What kinds of information would you be able to gather to help us with this?" The teacher was involved in identifying and analyzing the problem and in designing an intervention. Teachers in both the involvement and noninvolvement scenarios initially offered the same information, but in the teacher involvement condition, teacher input was encouraged and utilized by the psychologist in a collaborative process of analyzing difficulties and coming to some joint conclusions regarding an appropriate intervention. The psychologist in the videotape encouraged the teacher to express her concerns, ideas, observations, and intuitions regarding the problem situation. The psychologist guided the teacher in putting her ideas together and formulating a plan for intervention.

In the contrasting teacher involvement scenarios, a lack of teacher involvement was depicted as the consultant directly stating what should be done and telling the teacher how to do it. In the noninvolvement condition the discussion during problem identification was as follows: "During the next week, keep track of how many times you actually have to speak to Billy during math class for non- instructional reasons." The teacher shared information regarding the problem situation and received all recommendations from the school psychologist regarding an analysis of the difficulties and procedures for intervention. The school psychologist presented her perception and analysis of the problem and then explained the type of intervention that should be used. Thus, the teacher was essentially told what was wrong and what she should do. The school psychologist acted as an expert deciding what the problem was and how the teacher should deal with it.

Technical language versus ordinary language. Subjects viewed taped scenarios involving technical or nontechnical language. Within each condition, subjects viewed scenarios that either did or did not include teacher involvement. One group of subjects viewed a consultant using behavioral terminology to communicate with the teacher. In this scenario, the psychologist/consultant relied on the use of behavioral terms (e.g., reinforcement, contingencies, extinguish, conditioning, shaping, tokens, behavior modification) to appraise the situation and formulate a plan for intervention. For example, in the technical language condition the consultant stated during the problem identification phase: "So Pam, if you were to give a sequential analysis of these target behaviors, how would you describe them?"

Subjects in the nontechnical language viewed a consultant using nontechnical language or the teacher's own terminology to analyze classroom difficulties and formulate a plan. In the nontechnical variation the consultant said: "So Pam, if you were giving a step-by-step description of the actions you would like changed, how would it go?" Thus, the psychologist in this scenario used nontechnical language (e.g., praise, rewards, stop, change) rather than behavioral terms to make identical appraisals and recommendations of the same problem situation presented by the teacher.

Design. This study utilized a two-by-two factorial design with language and teacher involvement as the manipulated independent variables. The dependent variable, teachers' acceptability ratings of the consultation scenarios, was analyzed along three dimensions to include main effects for language, teacher involvement, and potential interactions.

Appendix F

Measure. After viewing videotaped scenarios, teachers were asked to complete the Intervention Rating Profile–15 (IRP–15), a measure designed to evaluate perceptions of acceptability (Martens et al., 1985). The IRP–15 is a global measure including 15 items rated on a 6-point Likert scale from "strongly disagree" to "strongly agree." Scores can range from 15 to 90 with higher scores indicating greater acceptability. The IRP–15 is composed of one primary factor, a general acceptability dimension reflecting the degree to which an intervention is judged to be suitable for use in regular classroom settings. The IRP–15 has a good psychometric foundation with a reliability of .98 using Cronbach's alpha (Martens et al., 1985). A factor analysis of the IRP–15 yielded one primary factor with item loadings greater than .82. The brief instructions to the IRP–15 were modified to reflect the purpose of the study, for example, to obtain information that would aid in the selection of styles of consultation and to evaluate consultation which is often provided to a teacher to help children with behavior problems. The only variation to the IRP–15 items made in the present study involved adding and/or substituting the term consultation for/with intervention.

Results

Subjects' acceptability ratings for the consultation scenarios were determined by total scores obtained on the IRP–15. Means and standard deviations for each condition are presented in Table 1. A two-way analysis of variance was completed with consultant language and consultee involvement as the independent variables and acceptability via the IRP–15 as the dependent variable. There was no main effect for involvement ($p > .05$) and no main effect for language ($p > .05$).

A significant interaction was obtained between consultant language and consultee involvement, $F(1,56) = 6.32, p.05$. Subjects gave the highest acceptability ratings to the scenario in which teacher involvement was low and the psychologist used technical language. Nontechnical language and low teacher involvement was rated as least acceptable. A

Table 1. Means and Standard Deviations for Acceptability Ratings

	Language	
Conditions	Technical	Nontechnical
Noninvolved		
Noninvolved M	73.47	58.40
Noninvolved SD	10.80	17.98
Involved		
Involved M	64.20	68.23
Involved SD	18.40	9.49

n = 15/cell.

Scheffe post hoc test indicated that the difference between the two language groups (Technical and Nontechnical) was relatively greater with the Low Involvement than with the High Involvement, with the net increase in the difference as a function of involvement being about 11 points, with a range (determined from a 95So CI) from 13.06 to 35.14.

Discussion

This study makes two contributions to the study of teacher reactions to behavioral consultation. First, this is the first time consultee involvement has been directly studied via acceptability ratings as a component of the consultation process. Previous researchers (Brigham & Stoerzinger, 1976; Elliott, 1988; Hughes & Falk, 1981; Kazdin, 1980) have suggested the potential importance of consultee involvement but none examined it directly, although our concept is similar to "collaboration" examined in a number of studies. Results of this study indicate that changes in the degree of teacher/consultee involvement or psychologist/consultant directiveness may, in fact, mediate the effect of certain other variables, such as consultant language. Researchers have often suggested that consultants engage in a collaborative relationship with the consultee by emphasizing joint problem solving (Hughes & Falk, 1981) and consumer's acceptability of behavioral interventions is enhanced when consultees are given a direct role in implementing and negotiating treatment (Brigham & Stoerzinger, 1976; Kazdin, 1980). The direct involvement of the consultee can also help ensure that an appropriate intervention is planned and that monitoring of the intervention is completed effectively (Clark, 1979; Gutkin & Curtis, 1990). In contrast, Erchul (1987) and Erchul and Chewning (1990) reported results that challenge the traditional concept of collaborative relationships, at least in behavioral consultation. The findings of the present study, that consultees do not show preference for teacher involvement necessarily, may be due to the language used and the measure of acceptability. Of course, researchers have used a variety of different measures across studies, thereby making results more difficult to compare.

Second, an analysis of scores obtained on the IRP–15 indicated that differences in consultant language (technical vs. nontechnical) did not cause acceptability ratings to differ significantly. It therefore appears that although the consultation research has identified consultant language as a variable that may affect consultation acceptability (e.g., Kazdin & Cole, 1981; Witt et al., 1984), a direct relationship may not exist. Researchers have advocated a softening of technical terminology (e.g., Kauffman & Hicente, 1972), but this was many years ago when behavioral techniques were still relatively new and, perhaps, more controversial. Over the years behavioral strategies have been refined, used effectively in the schools, incorporated in parent-training programs, and presented as a component of teacher training programs. There may now be less bias against technical terminology and behavioral techniques. It may now be more acceptable for consultants to achieve a mix of technical and nontechnical language so that the consultee perceives the consultant as an expert with knowledge to share. Unfortunately, in our study we did not obtain information on teacher skill, knowledge, and exposure to behavioral techniques. Future researchers should consider this assessment to extend the findings from our study.

In a study with undergraduate students, presenting behavioral treatments in technical rather than ordinary language was associated with more positive evaluations (Kazdin & Cole, 1981). The authors had not anticipated this result and suggested that the preference for technical language was related to the nature of the subject sample (e.g., college students) or to an increased respectability associated with technical language. In the current study, completed with videotaped scenarios and actual practicing teachers, a similar result was obtained. When the psychologist presented treatments in behavioral language, a more positive evaluation was observed if the teacher was the recipient of advice. Mean scores for acceptability indicated that technical language did receive a more favorable rating when subjects viewed the scenario in which the consultee was not involved. It may be that when the psychologist did not engage the teacher, the consultant became the pivotal point of attention and was, in fact, making a presentation, rather than engaging in a collaborative problem-solving process. If teachers did view the scenarios in which the consultee was not involved from this perspective, the consultant using technical terms may have seemed more knowledgeable, professional, and potentially effective.

In addition to enhancing the credibility and expertise of the consultant, the use of technical language could be beneficial in establishing the credibility of the intervention as a scientific technique. Informal interviews with subjects after the videotapes had been viewed and rated indicated that teachers wanted the psychologist to take a directive role. The implication was that if consultees knew how to handle the difficulty themselves they wouldn't need the consultant in the first place. Teachers also stated that they didn't object to the use of technical terms and that the implication that technical terminology could interfere with effective communication was distasteful in assuming that the teacher had a limited understanding of behavioral terms. This finding is consistent with previous work which indicates that although school psychologists preferred a mental health model of consultation, teachers preferred a behavioral approach largely because the consultant was perceived as specific, competent, efficient, and direct (Medway & Forman, 1980). With regard to generalizing results from the study, participants viewed and rated only one scenario and were unable to view or rate a contrast between the conditions of language and involvement. Because subjects had not viewed a scenario illustrating the contrasting condition (i.e., consultee noninvolvement), they were less able to rate the adequacy of the condition (i.e., consultee involvement) in the scenario they did view. An alternative strategy in future research would include using two scenarios and have subjects do a comparison rating or separate ratings with controls for ordering effects. Noninvolvement might also be further separated by phase of consultation. For example, the consultant directives regarding what should be done and how to do it might be separated in future research because they are related to different phases of behavioral consultation.

Although the use of a standard measure is desirable, the sole use of the IRP–15 may limit generalizations. The IRP–15 is a good measure of acceptability, but, it may be better in future research to use measures designed to assess the effects of language and involvement in addition to the IRP–15.

Finally, generalizability might be limited through the analogue characteristics of this research. For example, the consultation process was condensed to 12 minutes, the consultation was contrived, and teachers viewed a video rather than a real case. However, we believe that understanding of the variables manipulated in this study will be advanced through *both* analogue and naturalistic approaches (see Huebner, 1991, for a similar perspective in special education decision-making research). The tighter controls imposed on analogue research yield greater internal validity, but limit generalizations. Consistency of findings across analogue and naturalistic studies also yields stronger confidence in a research base. Obviously, the cost of naturalistic studies prohibits much needed research in this area.

References

Babcock, N. L., & Pryzwansky, W. B. (1983). Models of consultation: Preferences of educational professionals at five stages of service. *Journal of School Psychology, 21*, 359–366.

Bergan, J. R., & Kratochwill, T. R. (1990). *Behavioral consultation in applied settings.* New York: Plenum Press.

Brigham, T. A., & Stoerzinger, A. (1976). An experimental analysis of children's preference for self-selected rewards. In T. A. Brigham, R. P. Hawkins, J. Scott, & T. F. McLaughlin (Eds.), *Behavioral analysis in education: Self-control and reading* (pp. 51–63). Dubuque, IA: Kendall/Hunt.

Clark, R. D. (1979). Should consultants give teachers what they want—a straight answer? *Proceedings of the 11th Annual Convention of the National Association of School Psychologists,* p. 3.

Elliott, S. N. (1988). Acceptability of behavioral treatments in educational settings. In J. C. Witt, S. N. Elliott, & F. M. Gresham (Eds.), *The handbook of behavior therapy in education* (pp. 121–150). New York: Plenum.

Erchul, W. P. (1987). A relational communication analysis of control in school consultation. *Professional School Psychology, 2,* 113–124.

Erchul, W. P., & Chewning, T. G. (1990). Behavioral consultation from a request-centered relational communication perspective. *School Psychology Quarterly, 5,* 1–20.

Fairchild, T. N. (1976). School psychological services: An empirical comparison of two models. *Psychology in the Schools, 13,* 156–162.

Fine, M. J., Grantham, V. L., & Wright, J. G. (1979). Personal variables that facilitate or impede consultation. *Psychology in the Schools, 16,* 533–539.

Gutkin, T. B., & Curtis, M. (1990). School-based consultation: Theory, techniques, and research. In T. B. Gutkin & C. R. Reynolds (Eds.), *The handbook of school psychology* (2nd ed., pp. 577–634). New York: Wiley.

Huebner, E. S. (1991). Bias in special education decisions: The contribution of analogue research. *School Psychology Quarterly, 6,* 50–65.

Hughes, J. W., & Falk, R. S. (1981). Resistance, reactance and consultation. *Journal of School Psychology, 10,* 263–268.

Kauffman, J. M., & Hicente, A. P. (1972). Bringing in the sheaves: Observations on harvesting behavior change in the field. *Journal of School Psychology, 10,* 262–268.

Kazdin, A. E. (1980). Acceptability of time out from reinforcement procedures for disruptive child behavior. *Behavior Therapy, 11,* 329–344.

Kazdin, A. E., & Cole, P. M. (1981). Attitudes and labeling biases toward behavior modification: The effects of labels, content, and jargon. *Behavior Therapy, 12,* 56–68.

Martens, B. K., Peterson, R. L., Witt, J. C., & Cirone, S. (1986). Teacher perceptions of school-based intervention: Ratings of intervention effectiveness, ease of use, and frequency of use. *Exceptional Children, 53,* 213–223.

Martens, B. K., Witt, J. C., Elliott, S. N., & Darveaux, D. K. (1985). Teacher judgments concerning the acceptability of school-based interventions. *Professional Psychology: Research and Practice, 16,* 191–198.

Medway, F. J., & Forman, S. G. (1980). Psychologists' and teachers' reactions to mental health and behavioral school consultation. *Journal of School Psychology, 18,* 338–348.

Reimers, T. M., Wacker, D. P., & Koeppl, G. (1987). Acceptability of behavioral interventions: A review of the literature. *School Psychology Review, 16,* 212–227.

Reinking, R. H., Livesay, G., & Kohl, M. (1978). The effects of consultation style on consultee productivity. *American Journal of Community Psychology, 6,* 283–290.

Wenger, R. D. (1979). Teacher response to collaborative consultation. *Psychology in the Schools, 16,* 127–131.

Wiese, M. R., & Conoley, J. C. (1989). *The relationship of personal problem solving to consultees' preference for collaborative versus expert consultation.* Paper presented at the annual meeting of the American Psychological Association, New Orleans, LA.

Witt, J. C. (1990). Face-to-face verbal interaction in school-based consultation: A review of the literature. *School Psychology Quarterly, 5,* 199–210.

Witt, J. C., & Elliott, S. N. (1985). Acceptability of classroom intervention strategies. In T. R. Kratochwill (Ed.), *Advances in school psychology (Vol. IV,* pp. 251–288). Hillsdale, NJ: Erlbaum.

Witt, J. C., & Martens, B. K. (1983). Assessing the acceptability of behavioral interventions used in classrooms. *Psychology in the Schools, 20,* 510–517.

Witt, J. C., Moe, G., Gutkin, T. B., & Andrews, L. (1984). The effect of saying the same thing in different ways: The problem of language and jargon in school-based consultation. *Journal of School Psychology, 22,* 361–367.

Woolfolk, A. E., & Woolfolk, R. L. (1979). Modifying the effect of the behavior modification label. *Behavior Therapy, 10,* 575–578.

Woolfolk, A. E., Woolfolk, R. L., & Wilson, G. T. (1977). A rose by any other name. . . : Labeling bias and attitudes toward behavior modification. *Journal of Consulting and Clinical Psychology, 45,* 184–191.

Notes

1. Qualitative-Quantitative Research

1. The dramatic impact of the works of Deming (1991) and Barker (1992) to organizations exemplify strong paradigmatic shifts.

2. *Data* is a word we use in both quantitative and qualitative research to refer to the evidence (numerical evidence or narrative evidence). We use it because it is simple in form. We do acknowledge (and use from time to time), the term preferred by Denzin and Lincoln (1994), *empirical materials,* which is more accurate in qualitative research and may be more appealing to some readers.

3. On the issue of "truth," see, for example, Guba & Lincoln, 1982, 1985; Howe & Eisenhart, 1990; Kvale, 1983; LeCompte & Goetz, 1982; Miles & Huberman, 1984; J. K. Smith, 1983; and J. K. Smith & Heshusius, 1986.

2. Qualitative and Quantitative Research Methods

1. For a sampling of those researchers who apply multiple methods to their research, see Placek & Dobbs, 1988; Ragin, 1987; Reichardt & Cook, 1979; Shulman, 1986; and Stivers & Srinivasan, 1991.

2. For examples of studies in which the competitive basis of qualitative versus quantitative research is discussed, see Guba, 1978; Guba & Lincoln, 1982, 1989; Howe & Eisenhart, 1990; Kvale, 1983; LeCompte & Goetz, 1982; Miles & Huberman, 1984; J. K. Smith, 1983; and J. K. Smith & Heshusius, 1986.

3. Validity and Legitimation of Research

1. There are many measurement texts that provide detailed discussions (e.g., Ary et al., 1990; Gay, 1987; McMillan & James, 1992).

2. Higher-order factorial design is defined in a number of classical statistical textbooks (e.g., Edwards, 1960; Kirk, 1968). Discussion of these designs is beyond the scope of this book. See the glossary for definitions of multivariate and univariate analyses.

4. Strategies to Enhance Validity and Legitimation

1. Spradley (1979) describes elements of the ethnographic interview, including several types of questions: ethnographic, descriptive, structural, contrast, cultural-ignorance expression, repeating, restating, and so on (p. 67).

5. Applying the Qualitative-Quantitative Interactive Continuum to a Variety of Studies

1. The critique was completed by the spring 1991, University of Akron graduate research class: Sally Gartner, Miriam Keresman, Sandi Sommers, Jayne Speicher, and Brian Tindall.

Glossary

accretion measure. "An unobtrusive measure utilizing deposited physical material" (K. D. Bailey, 1978, p. 429; see also Webb et al., 1972).

case record. Condensation of raw data into a manageable, readable package.

case study. An in-depth study of all pertinent aspects of a person, thing, situation, institution, community, and so on (K. D. Bailey, 1978, p. 429; Good, 1963, p. 388; Mouly, 1970, p. 347).

classical theory building vs. grounded theory building. In *classical theory building*, one (1) establishes a concept or proposition; (2) develops hypotheses; and (3) conducts measurements and analyzes to verify the hypotheses. In *grounded theory building*, one (1) collects and analyzes data; (2) considers only those variables and hypotheses that emerge from the data; and (3) formulates a concept or proposition from the emergent relationships.

closed question. A questionnaire item in which optional response categories are provided for the respondent (K. D. Bailey, 1978, pp. 104, 430; Mouly, 1970, p. 249; Newman, 1976, p. 10).

comparability. The degree to which the ethnographer delineates the constructs generated and the characteristics of the groups studied so that they can be compared to other like and unlike groups (LeCompte & Goetz, 1982, p. 34; Wolcott, 1973).

concurrent validity. The estimate of how well a test correlates with another test that is considered to be valid.

construct validity. A conglomeration of all other types of validity. (Factor analysis is also frequently used to estimate construct validity.) The degree to which a test has construct validity is related to how well the test estimates the psychological construct being measured. Most frequently, authors refer to construct validity by summarizing several studies that give supportive evidence.

content validity. Also called *logical validity*; an estimate of how representative the test items are of the content or subject matter the test purports to measure. Frequently uses a table of specifications to help estimate the content representativeness.

criterion validity. Name given to predictive and concurrent validity taken together. This type of validity has also been called *empirical* or *statistical validity*.

deductive reasoning. Process that is part of the scientific way of knowing, whereby through a series of logical steps conclusions can be reached based on valid premises; usually considered to be the basis of quantitative research methods.

descriptive study. A study that uses statistical techniques to describe a sample (for example, by computing the mean) rather than to make inferences from the sample to the population or to use tests of statistical significance (Good, 1963, p. 242; Mouly, 1970, p. 234).

design validity. The extent to which the design is capable of answering the research question and/or the extent to which it can eliminate alternative explanations of the stated relationship (see *internal validity*). If the intent of the study is to generalize, then external-validity questions have to be answered to estimate the design validity of the study.

direct observation. Same as *participant observation* (Webb et al, 1972, p. 113).

epistemology. The study of the nature and grounds of knowledge.

ethnography or ethnomethodology. "Studying the commonsense features of everyday life, with emphasis on those things that 'everyone knows'" (K. D. Bailey, 1978, p. 249); social interaction usually is the focus of the study; it comes from traditions in anthropology and sociology (pp. 249–264, 432).

expert-judge validity. Content experts judge whether or not the test is measuring that which it purports to measure. Similar to *face validity*.

external criticism. The method by which historians determine the genuineness and authenticity of a document (Good, 1963, pp. 200–201).

external validity. The extent to which results of a study are generalizable to other people, groups, and so on (Newman & Newman, 1994, p. 229).

face validity. An estimate of participant reaction to the test. If the test appears to the person taking it to be measuring that which it purports to measure, then to that extent it has face validity.

field study. A research strategy in which ethnographic methods are used, and participant-observation strategies are usually employed in natural settings; can be a synonym for *ethnography* (K. D. Bailey, 1978, pp. 221, 433; Lofland, 1971).

focused coding. A part of the grounded theorists' *processual analysis,* which consists of taking limited sets of codes from the initial coding and applying them to large amounts of data; the level of coding consists of developing categories rather than simply labeling (Charmaz, 1983, p. 116; Glaser & Strauss, 1967).

foundational vs. antifoundational assumption. In a *foundational assumption,* reality can be "known" independent of the values of the "knower"; there is certitude and objectivity, as opposed to subjectivity, or "mind-dependent" reality of *antifoundational assumptions* (J. K. Smith, 1990).

grounded theory. A theory generated by or grounded in data rather than being abstract or tentative. Also refers to a research process stressing discovery and theory building through methods of initial and focused coding and memo writing (K. D. Bailey, 1978, p. 44; Charmaz, 1983; Glaser & Strauss, 1967).

indirect observation. A research strategy that is nonreactive, where observational data are collected by unobtrusive means (photographs or films, accretion and/or erosion measures) (K. D. Bailey, 1978, p. 239; Webb et al., 1972).

inductive reasoning. Reasoning from particular facts to a general conclusion; a process that is part of the scientific way of knowing (traditionally used by qualitative researchers) whereby observations or other bits of information (data) are collected, without preconceived notions of their relationships (hypotheses), with the assumption that relationships will become apparent, that conclusions will emerge from the data (Mouly, 1970, p. 30).

initial coding. Part of the grounded theorists' *processual analysis,* which consists of labeling descriptive information by code names (Charmaz, 1983, p. 113; Glaser & Strauss, 1967).

interactive continuum. A paradigm whereby two phenomena, while conceptually representing bipolar ends of a continuum, can also be used (to a greater or lesser degree in each) at any point along the continuum as the underlying sets of assumptions shift (i.e., interact); it is the descriptor of the research methodology and philosophy proposed in this book.

internal criticism. A method by which historians determine the meaning and trustworthiness of statements within a document (Good, 1963, p. 211).

internal validity. The degree to which all variables, except the one(s) under study, are controlled for in a research design (Keppel, 1973, p. 314; Newman, 1976, p. 231).

interview bias. The effect on the report of interviewer-collected data that comes from the personal attitudes, prejudices, and presuppositions of the interviewer.

interview schedule. A list of questions developed prior to an interview to be used by the interviewer (K. D. Bailey, 1978, p. 434).

isomorphism. Literally, "equal in forms." It is used in the philosophy of science to indicate that truth (knowledge) corresponds to reality (J. K. Smith, 1985).

known-group validity. A type of concurrent validity, it is an estimate of how well a test discriminates between identified groups.

measurement validity. The extent to which an instrument measures what it purports to measure.

memo writing. Part of the grounded theorists' *processual analysis* between coding and writing the results by which the researchers elaborate on the categories of data and the relationships among them. Memos are subsequently "sorted" and "integrated." Some researchers write many short memos on many categorical relationships, and over time the analytical level (of the ideas and the memos) becomes more abstract and, through this, theory is built (Charmaz, 1983, p. 120; Glaser & Strauss, 1967).

multiple regression analysis. Sometimes referred to as the general case of *least squares solution*, the basic underlying concept is that more than one predictor variable is used to predict one criterion variable; sometimes interchanged with the concept of *multiple correlation technique*.

multivariate. A condition in which there are two or more dependent (or criterion) variables being predicted by two or more independent (predictor) variables.

naturalistic inquiry. The type of research that is generally subsumed under qualitative methods; it is based on the underlying assumptions that knowledge about reality is mind dependent not value free; that hypotheses are always working hypotheses; that reality is not a single construct but multiple constructs. Techniques such as interview, observation, unobtrusive measurement, and document analysis are typically used to glean information in natural settings.

nonparticipant observation. A research situation in which the researcher collects data as an outsider and does not participate in ongoing activities (K. D. Bailey, 1978, p. 215).

nonstructured interview. A research strategy whereby the interviewer has a topic in mind but no predetermined questions (K. D. Bailey, 1978, p. 176; Newman, 1976, p. 12).

open-ended question. A type of questionnaire item for which predetermined response options are not provided (K. D. Bailey, 1978, pp. 104, 435; Mouly, 1970, p. 249; Newman, 1976, p. 9).

partially structured interview. A research strategy whereby some questions are predetermined and the interviewer also uses open-ended questions and probes to explore more in-depth reasons for answers.

participant observation. A research situation in which the researcher is a regular participant in the activities being observed while he/she collects data; the dual role is usually not known to the other participants (K. D. Bailey, 1978, p. 215, 435; Webb et al., 1972, p. 115).

phenomenological research. A qualitative research method founded by Edmund Husserl as a reaction against the empiricist conception of the world as an "objective universe of facts" (Kvale, 1983). The researcher is involved in three aspects: "open description," "investigation of essences," and "phenomenological reduction," whereby interview responses are recorded, transcribed, and reviewed for central themes of meaning (Mitchell, 1990).

predictive validity. The estimate of how well a test predicts eventual outcome.

primary source. A document or data set provided by actual witnesses to the incident or phenomenon under study (Good, 1963, p. 194; Mouly, 1970, p. 213).

processual analysis. A research method used by grounded theorists in which the process of coding the data constitutes the analysis of the data (Charmaz, 1983, p. 117; Glaser & Strauss, 1967).

qualitative analysis. Data analysis in which the aim is building theory. It is generally inductive in approach, is based originally on the naturalistic assumption that reality is mind dependent (i.e., can only be known as it is interpreted and has "meaning" for the observer), is usually of single-subject design, and generally deals with nominal data (K. D. Bailey, 1978, p. 436; LeCompte & Goetz, 1982; Reichardt & Cook, 1979, p. 10; Rist, 1977; J. K. Smith, 1983).

quantitative analysis. Data analysis in which the aim is theory testing. It is generally deductive in approach, is based originally on the rationalistic assumption that reality is mind independent (i.e., can separate the observer from the object of study), and has as its goal generalizability. It usually deals with ordinal, interval, or ratio data (K. D. Bailey, 1978, p. 436; LeCompte & Goetz, 1982; Reichardt & Cook, 1979, p. 10; Rist, 1977; J. K. Smith, 1983).

rationalist vs. naturalist. *Rationalist* philosophy holds that one can "know" reality as objective phenomena, outside the influence of the "knower" and his/her values; whereas *naturalist* philosophy holds that one cannot separate the "known" reality from the values of the "knower";

in other words, knowledge that is mind independent vs. knowledge that is mind dependent (Mouly, 1970; J. K. Smith, 1983, 1985).

raw case data. All information collected about a research topic (person, place, or thing).

reliability. A value indicating the internal consistency of a measure or the repeatability of a measure or finding; the extent to which a result or measurement will be the same value every time it is measured (Keppel, 1973, p. 310; Newman & Newman, 1994).

scientific method. The step-by-step process by which theory is both generated and verified; both an inductive and a deductive process. Generally a phenomenon is observed; set(s) of relationships is (are) stated; a hypothesis is stated; a design is created to test the hypothesis; data are collected; data are analyzed; the results are concluded; the hypothesis is verified or refuted; theory is refined (Good, 1963, p. 5).

secondary source. A source containing an intermediate person's (not the actual witnesses') reporting of the event or phenomenon under study (Good, 1963, p. 194; Mouly, 1970, p. 213).

sorting memos. Part of the grounded theorists' processual analysis. It follows initial coding, focused coding, and writing memos; the researcher analyzes relationships among memos and pulls them together to develop theoretical relationships (Charmaz, 1983).

structured interview. A research strategy in which the interviewer reads each question and the possible answers; the respondent responds; and the response is recorded (K. D. Bailey, 1978, p. 170; Newman, 1976, p. 12).

structured observation. A research situation in which data are collected by the observer's use of predetermined categories of behaviors (K. D. Bailey, 1978, pp. 216, 231).

synthesis. The process of blending external criticism and internal criticism to report historical data accurately (Mouly, 1970, p. 211).

theoretical sampling. The sampling of additional data to develop an emerging theory (Charmaz, 1983 p. 124–125). "The process of data collection for generating theory whereby the analyst jointly collects, codes, and analyzes his data and decides what data to collect next and where to find them, in order to develop his theory as it emerges" (Glaser & Strauss, 1967, p. 45).

theory building vs. theory testing. The difference some use to dichotomize qualitative and quantitative philosophies and methods; namely, that qualitative methodologists collect data with neither a theoretical base nor a hypothesis and use the data to generate categories as well as statements of relationships, *theory building*; while quantitative investigators begin with a theoretically based hypothesis and collect previously established categories of data to test the viability of that hypothesis, *theory testing*.

translatability. The degree to which the research methods, analytical categories, and characteristics of phenomena and groups are explicitly described by ethnographers so that comparisons with other groups can be made (LeCompte & Goetz, 1982, p. 34).

triangulation. The combining of two or more data-collection methods and/or data sources into one design (K. D. Bailey, 1978, p. 239; Jick, 1979; LeCompte & Goetz, 1982, p. 35; Webb et al., 1972).

univariate. One variable (dependent or criterion variable) is being predicted by another set of variables (independent or predictor variable). This set of predictor variables could be one or more variables.

unobtrusive measure. A nonreactive measure in which the behavior of the subjects being studied is not changed because they do not know research is being conducted (K. D. Bailey, 1978, p. 239; Webb et al., 1972).

unstructured observation. A research strategy in which observational data are collected without predetermined categories to look for or hypotheses to guide the observation (K. D. Bailey, 1978, p. 216).

validity. The degree to which one actually is measuring what one wishes to measure; several types exist (Keppel, 1973, p. 310; Mouly, 1970, p. 118; Newman, 1976, pp. 56, 240).

verstehen. The researcher attempts to portray the meaning of the lives of those he or she studies from their (those being studied) point of view.

Bibliography

Alexander, J. C., & Harman, R. L. (1988). One counselor's intervention in the aftermath of a middle school student's suicide: A case study. *Journal of Counseling and Development, 66*, 283–285.

Ary, D., Jacobs, L. C., & Razavieh, A. (1990). *Introduction to research in education*. Orlando, FL: Holt, Rinehart & Winston.

Bailey, J. T. (1956). The critical incident technique in identifying behavioral criteria for professional nursing effectiveness. *Nursing Research, 5*(2), 54–58.

Bailey, K. D. (1978). *Methods of social research*. New York: Macmillan.

Barker, J. (1992). *Future edge*. New York: Morrow.

Becker, H. S., & Geer, B. (1960). Participant observation: The analysis of qualitative field data. In R. Adams & J. Preiss (Eds.), *Human organizational research*. Homewood, IL: Dorsey Press.

Beld, J. M. (1994). Constructing a collaboration: A conversation with E. G. Guba and Y. S. Lincoln. *Qualitative Studies in Education, 7*(2), 99–115.

Benz, C., & Newman, I. (1986). *Qualitative-quantitative interactive continuum: A model and application to teacher education evaluation*. Paper presented at the American Association of Colleges for Teacher Education meeting, Chicago, Il. (ERIC Document Reproduction Service No. ED 269 406)

Blumer, H. (1980). Comment, Mead and Blumer: The convergent methodological perspective of social behaviorism and symbolic interaction. *American Sociological Review, 45*, 409–419.

Bogdan, R. C., & Biklen, S. K. (1982). *Qualitative research for education: An introduction to theory and methods*. Boston: Allyn & Bacon.

Boostrom, R. (1994). Learning to pay attention. *Qualitative Studies in Education, 7*(1), 51–64.

Borg, W. R., & Gall, M. D. (1989). *Educational research: An introduction*. New York: Longman.

Boyan, N. J. (Ed.). (1988). *Handbook of research on educational administration*. New York: Longman.

Bibliography

Campbell, D. T., & Fiske, D. W. (1959). Convergent and discriminant validation by the multitrait-multimethod matrix. *Psychological Bulletin, 56,* 81–105.

Campbell, D. T., & Stanley, J. C. (1963). *Experimental and quasi-experimental designs for research.* Chicago: Rand McNally.

Charmaz, K. (1983). Grounded theory method: An explication and interpretation. In R. Emerson (Ed.), *Contemporary field research* (pp. 109–126). Boston: Little, Brown.

Comte, A. (1974). *Discours sur l'esprit positif.* Paris: Librairie Philosophique. (Original work published in 1844)

Cook, T. D., & Reichardt, C. S. (1979). *Qualitative and quantitative methods in evaluation research.* Beverly Hills, CA: Sage.

Creswell, J. W. (1944). *Research design: Qualitative and quantitative approaches.* Thousand Oaks, CA: Sage.

Culbertson, J. A. (1988). A century's quest for a knowledge base. In N. J. Boyan (Ed.), *Handbook of research on educational administration* (pp. 3–26). New York: Longman.

Curtis, J. M. (1981). Effect of therapist's self-disclosure on patients' impressions of empathy, competence, and trust in an analogue of a psychotherapeutic interaction. *Psychological Reports, 48,* 127–136.

Deming, W. E. (1991). *Out of the crisis.* Cambridge, MA: MIT Press.

Denzin, N. K. (1978a). The logic of naturalistic inquiry. In N. K. Denzin (Ed.), *Sociological methods: A sourcebook.* New York: McGraw-Hill.

Denzin, N. K. (1978b). *The research act* (2nd ed.). New York: McGraw-Hill.

Denzin, N. K. (1988). *The research act* (Rev. ed.). New York: McGraw-Hill.

Denzin, N. K. (1989). *The research act* (3rd ed.). Englewood Cliffs, NJ: Prentice-Hall.

Denzin, N. K. (1994). Evaluating qualitative research in the poststructural moment: The lessons James Joyce teaches us. *Qualitative Studies in Education, 7*(4), 295–308.

Denzin, N. K., & Lincoln, Y. S. (Eds.). (1984). *Handbook of qualitative research.* Thousand Oaks, CA: Sage.

Dewey, J. (1929). *The sources of a science of education.* New York: Liveright.

Diesing, P. (1991). *How does social science work? Reflections on practice.* Pittsburgh, PA: University of Pittsburgh Press.

Donmoyer, R. (1990). Generalizability and the single-case study. In E. Eisner & A. Peshkin (Eds.), *Qualitative inquiry in education: The continuing debate* (pp. 175–200). New York: Teachers College Press.

Douglas, J. (1976). *Investigative social research*. Beverly Hills, CA: Sage.

Dudley, N. Q. (1992). Participatory principles in human inquiry: Ethics and methods in a study of the paradigm shift experience. *Qualitative Studies in Education, 5*(4), 325–344.

Edwards, A. L. (1960). *Experimental design in psychological research*. New York: Holt, Rinehart & Winston.

Eisner, E. W. (1991). *The enlightened eye: Qualitative inquiry in the enhancement of educational practice*. New York: Macmillan.

Eisner, E. W., & Peshkin, A. (Eds.). (1990). *Qualitative inquiry in education: The continuing debate*. New York: Teachers College Press.

Erickson, F. (1986). Qualitative methods in research on teaching. In M. C. Wittrock (Ed.), *Handbook of research on teaching* (3rd ed., pp. 119–161). New York: Macmillan.

Fetterman, D. M. (1989). *Ethnography: Step by step*. Newbury Park, CA: Sage.

Filstead, W. J. (Ed.). (1970). *Qualitative methodology: First-hand involvement with the social world*. Chicago: Markham.

Firestone, W. (1987). Meaning in method: The rhetoric of quantitative and qualitative research. *Educational Researcher, 16*, 16–21.

Foerstner, S. B., Newman, I., & Koenig, D. (1985, October). *To laugh or not to laugh: Why? Its measurement and meaning*. Paper presented at the meeting of the Midwestern Educational Research Association, Chicago, IL.

Fraser, B., & Tobin, K. (1989, March). *Combining qualitative methods in the study of classroom learning environment*. Paper presented at the meeting of the American Educational Research Association, San Francisco, CA.

Fuller, M. L. (1994). The monocultural graduate in the multicultural environment: A challenge for teacher educators. *Journal of Teacher Education, 45*(4), 269–277.

Gay, L. R. (1987). *Educational research: Competencies for analysis and application* (3rd ed.). Columbus, OH: Merrill.

Geertz, C. (1973). Thick descriptors: Toward an interpretive theory of culture. In *The interpretation of culture*. New York: Basic Books.

Glaser, B. G. (1978). *Theoretical sensitivity*. Mill Valley, CA: Sociological Press.

Bibliography

Glaser, B. G. & Strauss, A. L. (1967). *The discovery of grounded theory: Strategies for qualitative research.* Chicago: Aldine Press.

Glass, G. V. (1971). *The growth of evaluation methodology* (AERA Curriculum Evaluation Monograph Series No. 7.). Chicago: Rand McNally.

Glesne, C., & Peshkin, A. (1992). *Becoming qualitative researchers: An introduction.* New York: Longman.

Goetz, J., & LeCompte, M. (1984). *Ethnography and qualitative design in educational research.* San Diego, CA: Academic Press.

Good, C. V. (1963). *Introduction to educational research.* New York: Appleton-Century-Crofts.

Guba, E. G. (1978). *Toward a methodology of naturalistic inquiry in educational evaluation* (CSE Monograph Series in Evaluation No. 8.). Los Angeles: Center for Study of Evaluation, University of California.

Guba, E. G. (Ed.). (1990). *The paradigm dialog.* Newbury Park, CA: Sage.

Guba, E. G., & Lincoln, Y. S. (1982). Epistemological and methodological bases of naturalistic inquiry. *Educational Communications and Technology Journal, 4*(30).

Guba, E. G., & Lincoln, Y. S. (1985, October). *The countenances of fourth generation evaluation: Description, judgment and negotiation.* Paper presented at the meeting of the Evaluation Network, Toronto, Canada.

Guba, E. G., & Lincoln, Y. S. (1989). *Fourth generation evaluation.* Newbury Park, CA: Sage.

Gumperz, J. (1986). Interactional sociolinguistics in the study of schooling. In J. Cook-Gumperz (Ed.), *The social construction of literacy* (pp. 45–68). Cambridge, MA: Harvard University Press.

Hakim, C. (1987). *Research design: Strategies and choices in the design of social research.* Boston: Allen & Unwin.

Hammersley, M. (1992). Some reflections on ethnography and validity. *Qualitative Studies in Education, 5*(3), 195–203.

Hempel, C. (1965). *Aspects of scientific explanation.* New York: Free Press.

Hillway, T. (1969). *Handbook of educational research.* Boston: Houghton-Mifflin.

Holland, D. C., & Eisenhart, M. A. (1990). *Educated in romance: Women, achievement, and college culture.* Chicago: University of Chicago Press.

Howard, D. C. P. (1994). Human-computer interactions: A phenomenological examination of the adult first-time computer experience. *International Journal of Qualitative Studies in Education* 7(1), 33–50.

Howe, K., & Eisenhart, M. (1990). Standards for qualitative (and quantitative) research: A prolegomenon. *Educational Researcher,* 19(4), 2–9.

Ianni, F. A., & Storey, E. (Eds.). (1973). *Cultural relevance and educational issues.* Boston: Little, Brown.

Jackson, P. (1968). *Life in classrooms.* New York: Holt, Rinehart & Winston.

Jick, T. D. (1979). Mixing qualitative and quantitative methods: Triangulation in action. *Administrative Science Quarterly, 24,* 602–611.

Keller, E. F. (1985). *Reflections on gender and science.* New Haven, CT: Yale University Press.

Keppel, G. (1973). *Design and analysis: A researcher's handbook.* Englewood Cliffs, NJ: Prentice-Hall.

Kerlinger, F. (1964). *Foundations of behavioral research.* New York: Holt, Rinehart & Winston.

Khleif, B. B. (1974). Issues in anthropological fieldwork in schools. In G. D. Spindler (Ed.), *Education and cultural process: Toward an anthropology of education* (pp. 389–398). New York: Holt, Rinehart & Winston.

Kirk, J., & Miller, M. L. (1985). *University Paper Series on Qualitative Research Methods, Vol. 1. Reliability and validity in qualitative research.* Beverly Hills, CA: Sage.

Kirk, R. E. (1968). *Experimental design: Procedures for the behavioral sciences.* Belmont, CA: Brooks.

Kitzinger, C. (1990). *The social construction of lesbianism.* Beverly Hills, CA: Sage.

Kuhn, T. S. (1970). *The structure of scientific revolutions* (2nd ed.). Chicago: University of Chicago Press.

Kvale, S. (1983). The qualitative research interview: A phenomenological and a hermeneutical mode of understanding. *Journal of Phenomenological Psychology, 14*(2), 171–195.

Kvale, S. (1995). The social construction of validity. *Qualitative Inquiry, 1*(1), 19–40.

Lather, P. (1986). Issues of validity in openly ideological research. *Interchange, 17*(4), 63–84.

Bibliography

Lather, P. (1995). The validity of angels: Interpretive and textual strategies in researching the lives of women with HIV/AIDS. *Qualitative Inquiry, 1*(1), 41–68.

LeCompte, M. D., & Goetz, J. P. (1982). Problems of reliability and validity in ethnographic research. *Review of Educational Research, 52*(1), 31–60.

Lincoln, Y. S. (1990). Toward a categorical imperative for qualitative research. In E. W. Eisne & A. Peshkin (Eds.), *Qualitative inquiry in education: The continuing debate* (pp. 277–295). New York: Teachers College Press.

Lincoln, Y. S. (1995). Emerging criteria for quality in qualitative research. *Qualitative Inquiry, 1*(3), 275–289.

Lincoln, Y. S., & Guba, E. G. (1985). *Naturalistic inquiry.* Beverly Hills, CA: Sage.

Lipman-Blumen, J. (1985). The creative tension between liberal arts and specialization. *Liberal Education, 71*(1), 18.

Lofland, J. (1971). *Analyzing social settings: A guide to qualitative observation and analysis.* Belmont, CA: Wadsworth.

Lofland, J. (1974). Style of reporting qualitative field research. *American Sociologist, 9*, 101–111.

Lombard, G. J. (1991, April). *Closing the loop: Adding the qualitative dimension to critical incidents.* Paper presented at the meeting of the American Educational Research Association, Chicago, IL.

Lyotard, J. F. (1984). *The postmodern condition: A report on knowledge.* Manchester, UK: Manchester University Press.

Marshall, C., & Rossman, G. B. (1989). *Designing qualitative research.* Newbury Park, CA: Sage.

Matz, J. L. (1977). *Interviewing: Principles, techniques, and applications.* San Antonio, TX: St. Mary's University.

McAshan, H. H. (1963). *Elements of educational research.* New York: Mc-Graw-Hill.

McMillan, H., & James, H. (1992). *Educational research: Fundamentals for the consumers.* New York : Harper Collins.

Merriam, S. B. (1988). *Case study research in education: A qualitative approach.* San Francisco: Jossey-Bass.

Miles, M. B., & Huberman, A. M. (1984). Drawing valid meaning from qualitative data: Toward a shared craft. *Educational Researcher, 13*(5), 20–30.

Miles, M. B., & Huberman, A. M. (1994). *Qualitative data analysis: An expanded sourcebook* (2nd ed.). Thousand Oaks, CA: Sage.

Miller, L., & Lieberman, A. (1988). School improvement in the United States: Nuance and numbers. *Qualitative Studies in Education, 1*(1), 3–19.

Miller, S. E., Leinhart, G., & Zigmond, N. (1988). Influencing engagement through accommodation: An ethnographic study of at-risk students. *American Educational Research Journal, 25,* 465–487.

Mitchell, J. G. (1990). *Re-visioning educational leadership: A phenomenological approach.* New York: Garland.

Mouly, G. J. (1970). *The science of educational research.* New York: Litton Educational.

Newman, I. (1976). *Basic procedures in conducting survey research.* Akron, OH: University of Akron.

Newman, I., & Newman, C. (1994). *Conceptual statistics for beginners.* Washington, DC: University Press of America.

Ödman, P. J. (1992). Interpreting the past. *Qualitative Studies in Education, 5*(2), 167–184.

Ogbu, J. (1987). Variability in minority response to schooling: Nonimmigrants vs. immigrants. In G. Spindler & L. Spindler (Eds.), *Interpretive ethnography of education: At home and abroad* (pp. 255–278). Hillsdale, NJ: Lawrence Erlbaum.

Olson, L. (1986, January 15). "Positive science" or "normative principles"? *Educational Week, 5*(18).

Patton, M. Q. (1980). *Qualitative evaluation methods.* Beverly Hills, CA: Sage.

Patton, M. Q. (1987). *Creative evaluation* (2nd ed.). Beverly Hills, CA: Sage.

Patton, M. Q. (1990). *Qualitative evaluation and research methods* (2nd ed.). Newbury Park, CA: Sage Publications.

Phillips, D. C. (1990). Subjectivity and objectivity: An objective inquiry. In E. W. Eisner & A. Peshkin (Eds.), *Qualitative inquiry in education: The continuing debate* (pp. 19–37). New York: Teachers College Press.

Placek, J. H., & Dobbs, P. (1988). A critical incident study of preservice teachers' beliefs about teaching success and nonsuccess. *Research Quarterly for Exercise and Sport, 59*(4), 351–358.

Pohland, P. (1972). Participant observation as a research methodology. *Studies in Art Education, 13*(3), 4–115.

Bibliography

Polkinghorne, D. E. (1983). *Methodology for the human sciences: Systems of inquiry.* Albany: State University of New York Press.

Polkinghorne, D. E. (1991, April). *Generalization and qualitative research: Issues of external validity.* Paper presented at the meeting of the American Educational Research Association, Chicago, IL.

Popper, K., (1962). *Conjectures and Refutations.* New York: Harper.

Ragin, C. C. (1987). *The comparative methods: Moving beyond qualitative and quantitative strategies.* Berkeley: University of California Press.

Reichardt, C. S., & Cook, T. D. (1979). Beyond qualitative *versus* quantitative methods. In T. D. Cook & C. S. Reichardt (Eds.), *Qualitative and quantitative methods in evaluation research* (pp. 7–32). Beverly Hills, CA: Sage.

Rhoades, M. M., & Kratochwill, T. R. (1992). Teacher reactions to behavioral consultation: An analysis of language and involvement. *School Psychology Quarterly, 7*(1), 47–59.

Richardson, S. A., Dohrenwend, B. S., & Klein, D. (1965). *Interviewing: Its forms and functions.* New York: Basic Books.

Ricoeur, P. (1988). *From text to action: An anthology on hermeneutics.* Stockholm, Sweden: Symposium.

Rist, R. C. (1977). On the relations among educational research paradigms: From disdain to détente. *Anthropology and Educational Quarterly, 8,* 42–49.

Robinson, V. J. (1992). Doing critical social science: Dilemmas of control. *Qualitative Studies in Education, 5*(4), 345–359.

Rosenthal, R., & Rosnow, R. L. (1991). *Essentials of behavioral research: Methods of data analysis.* New York: McGraw-Hill.

Rossman, G. B., Corbett, H. D., Firestone, W. A. (1988). *Change and effectiveness in schools: A cultural perspective.* Albany: State University of New York Press.

Schatzman, L., & Strauss, A. L. (1973). *Field research: Strategies for a natural sociology.* Englewood Cliffs, NJ: Prentice-Hall.

Schratz, M. (Ed.). (1993). *Qualitative voices in educational research.* London: Falmer Press.

Shafer, R. J. (1974). *A guide to historical method.* Homewood, IL: Dorsey Press.

Shaker, P. (1990). The metaphorical journey of evaluation theory. *Qualitative Studies in Education, 3*(4), 355–363.

Sherman, R. R., & Webb, R. B. (Eds.). (1988). *Qualitative research in education: Focus and methods.* London: Falmer Press.

Shulman, L. S. (1986). Paradigms and research programs in the study of teaching: A contemporary perspective. In M. C. Wittrock (Ed.), *Handbook of research on teaching* (3rd ed., pp. 3–36). New York: Macmillan.

Sindell, P. S. (1969). Anthropological approaches to the study of education. *Review of Educational Research, 39,* 593–605.

Smith, J. K. (1983). Quantitative versus qualitative research: An attempt to clarify the issue. *Educational Researcher, 12(*3), 6–13.

Smith, J. K. (1984). The problem of criteria for judging interpretive inquiry. *Educational Evaluation and Policy Analysis, 6*(4), 379–391.

Smith, J. K. (1985, April). *Closing down the conversation: The end of the qualitative-quantitative debates in educational inquiry.* Paper presented at the meeting of the American Educational Research Association, Chicago, IL.

Smith, J. K. (1990). Alternative research paradigms and the problem of criteria. In E. Guba (Ed.), *The paradigm dialog* (pp. 167–187). Newbury Park, CA: Sage.

Smith, J. K., & Heshusius, L. (1986). Closing down the conversation: The end of the quantitative-qualitative debate among educational researchers. *Educational Researcher, 15*(1), 4–12.

Smith, L. M. (1967). The microethnography of the classroom. *Psychology in the Schools, 4,* 216–221.

Smith, L. M., & Geoffrey, W. (1968). *Complexities of an urban classroom.* New York: Holt, Rinehart & Winston.

Spencer, H. (1910). *Essays: Scientific, political, and speculative.* (Vol. 2). New York: Appleton.

Spindler, G. D. (1974). *Educational and cultural process: Toward an anthropology of education.* New York: Holt, Rinehart & Winston.

Spindler, G. D. (Ed.). (1982). *Doing the ethnography of schooling.* Prospect Heights, IL: Waveland Press.

Spradley, J. P. (1979). *The ethnographic interview.* New York: Holt, Rinehart & Winston.

Bibliography

Stake, R. E. (1978, February). The case study method in social inquiry. *Educational Researcher, 7,* 5–8.

Stake, R. E. (1981). Case study methodology: An epistemological advocacy. In W. W. Welch (Ed.), *Case study methodology in educational evaluation: Proceedings of the 1981 Minnesota Evaluation Conference* (pp. 31–40). Minneapolis: Minnesota Research and Evaluation Center.

Stivers, E., & Srinivasan, R. (1991, April). *Using mathematics in qualitative research.* Paper presented at the meeting of the American Educational Research Association, Chicago, IL.

Strauss, A. L. (1987). *Qualitative analysis for social scientists.* Cambridge, MA: Harvard University Press.

Tierney, W. G. (1993). The cedar closet. *Qualitative Studies in Education, 6*(4), 303–314.

Trankell, A. (1972). *Reliability of evidence.* Stockholm, Sweden: Beckmans.

Trueba, H. T. (1991, April). *Back to basics in educational research: The uses and abuses of "culture" and "ethnography" in educational research.* Paper presented at the meeting of the American Educational Research Association, Chicago, IL.

Trueba, H. T., Jacobs, L., & Kirton, E. (1990). *Cultural conflict and adaptation: The case of Hmong children in American society.* London: Falmer Press.

Van Maanen, J. (1988). *Tales of the field: On writing ethnography.* Chicago: University of Chicago Press.

Van Manen, M. (1984). *Doing phenomenological research and writing: An introduction* (Curriculum Praxis monograph series No. 7). Edmonton: University of Alberta.

Van Manen, M. (1990). *Researching lived experience: Human science for an action sensitive pedagogy.* London, Ontario: Althouse Press.

Vidich, A. J., & Lyman, S. M. (1994). Qualitative methods: Their history in sociology and anthropology. In N. K. Denzin & Y. S. Lincoln (Eds.), *Handbook of qualitative research* (pp. 23–59). Thousand Oaks, CA: Sage.

Wax, M. L., Diamond, S., & Gearing, F. O. (Eds.). (1971). *Anthropological perspectives on education.* New York: Basic Books.

Webb, E. J., Campbell, D. T., Schwartz, R. D., & Sechrest, L. (1972). *Unobtrusive measures: Nonreactive research in the social sciences.* Chicago: Rand McNally.

Bibliography

Welch, W. W. (Ed.). (1981). *Case study methodology in educational evaluation: Proceedings of the 1981 Minnesota Evaluation Conference*. Minneapolis: Minnesota Research and Evaluation Center.

Wertz, F. J. (1986). The question of reliability of psychological research. *Journal of Phenomenological Psychology, 17*, 81–205.

Wiersma, W. (1980). *Research methods in education*. Itasca, IL: F. E. Peacock.

Wilson, S. (1977). The use of ethnographic techniques in educational research. *Review of Educational Research, 47*(1), 245–265.

Wittrock, M. C. (Ed.). (1986). *Handbook of research on teaching* (3rd ed.). New York: Macmillan.

Wolcott, H. F. (1973). *The man in the principal's office: An ethnography*. New York: Holt, Rinehart & Winston.

Wolcott, H. F. (1990a). On seeking—and rejecting—validity in qualitative research. In E. W. Eisner & A. Peshkin (Eds.), *Qualitative inquiry in education* (pp. 121–152). New York: Teachers College Press.

Wolcott, H. F. (1990b). *Writing up qualitative research*. Beverly Hills, CA: Sage.

Wolcott, H. F. (1994). *Transforming qualitative data: Description, analysis, and interpretation*. Thousand Oaks, CA: Sage.

Yarbrough, P. (1982). An ethnography of physical therapy practice: A source for curriculum development (Doctoral dissertation, Georgia State University, 1980). *Dissertation Abstracts International, 41*, 1914A.

Index

Index

Einstein, Albert, 6
Eisner, E. W., 3, 28
empirical studies, 10
Enlightened Eye, The (Eisner), 3
equivalent forms, 35
Erickson, Frederick, 24
Ethnographic Interview, The (Spradley), 68
ethnographic methods, 25, 76–80, 102–3; culture shock and, 77; internal validity and, 78–79
experimental mortality, 40
expert-judge content validity, 38
ex post facto research, 36, 41–43
external validity, 33, 35–36, 72–73; ex post facto research and, 43; multivariate research and, 44–45; published research and, 99, 107–8

face validity, 38, 58
factor analysis, 39
factorial designs, 44
falsifiability, 13
feedback loops, 22, 25–26, 111
Firestone, William, 2–3
Fisk, D. W., 83
Fuller, Mary Lou, 102–6

Gall, M.D., 41
Gay, L. R., 58
Gearing, Fred, 76
Geer, B., 59–60
Geertz, C., 2
generalizability: aggregate, 54; applicability, 54–55, 96, 105; consistency-questions model and, 89; context limited (transferability), 54–55, 96, 105; controls and, 36; ethnographic methods and, 79; replicability, 54–55, 96, 105; research design and, 54–55, 99–100; statistical, 54; threats to, 33–34
Glaser, B. G., 10, 61
Glesne, C., 17
Goetz, J., 34, 50, 78–79, 85, 110
Good, C. V., 70–71, 73
graduate programs, 7–8, 76–77
grounded theory methods, 17; initial codes, 61; memo writing, 64; theoretical sampling, 64–65
Guba, Egon, 18, 27, 32, 45, 50, 59–60, 112

Hakim, C., 68
Hammersley, M., 32, 46
Handbook of Qualitative Research (Denzin and Lincoln), 16
Handbook of Research on Teaching (Shulman), 109
Harman, R. L., 93, 97
historical methods, 70–76; causation and, 74; external and internal criticism, 72–73; scientific method and, 70–71; synthesis, 73
history, 37
history effects, 85
Howard, Dale, 81
Husserl, Edmund, 82
hypotheses, 98; derivation of, 25; grounded theory and, 61, 63; historical methods and, 72; observational methods and, 60; working, 24
hypothesis-testing research. *See* quantitative research

Index

Isadore Newman, a licensed psychologist, has taught at the University of Akron's College of Education for twenty-seven years. He is associate director of the Institute of Life Span Development and Gerontology; an adjunct professor, Department of Psychiatry, Division of Psychology, Northeastern Ohio Universities College of Medicine; and a distinguished Harrington professor. He has presented hundreds of refereed papers, written more than one hundred refereed articles, and authored or coauthored nine books and monographs. He has served on a number of editorial boards of research journals and holds leadership positions in several professional research organizations.

Carolyn R. Benz is a professor of educational administration at the University of Dayton, where she has been a faculty member since 1990. Her areas of expertise are program evaluation and research methodology. She regularly teaches qualitative and quantitative research methodology courses to doctoral students, attempting to maintain the interactive continuum that embraces both. She also teaches program evaluation and dissertation-proposal-writing courses, as well as directing doctoral dissertations. Benz has coauthored the recently published book *Getting Started in Writing the Behavioral Dissertation* with Isadore Newman, David Weis, and Keith McNeil. Among her publications are those in *American Educational Research Journal, Review of Higher Education, Journal of Educational Research,* and *Journal of Teacher Education.*

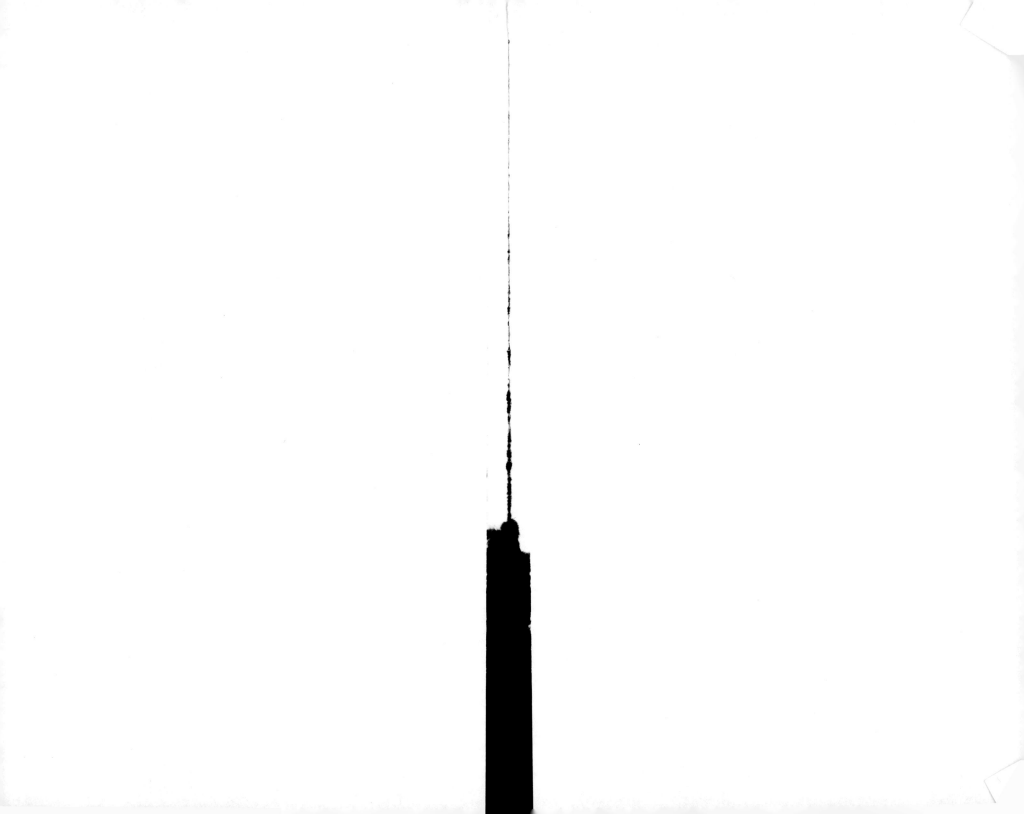

Isadore Newman, a licensed psychologist, has taught at the University of Akron's College of Education for twenty-seven years. He is associate director of the Institute of Life Span Development and Gerontology; an adjunct professor, Department of Psychiatry, Division of Psychology, Northeastern Ohio Universities College of Medicine; and a distinguished Harrington professor. He has presented hundreds of refereed papers, written more than one hundred refereed articles, and authored or coauthored nine books and monographs. He has served on a number of editorial boards of research journals and holds leadership positions in several professional research organizations.

Carolyn R. Benz is a professor of educational administration at the University of Dayton, where she has been a faculty member since 1990. Her areas of expertise are program evaluation and research methodology. She regularly teaches qualitative and quantitative research methodology courses to doctoral students, attempting to maintain the interactive continuum that embraces both. She also teaches program evaluation and dissertation-proposal-writing courses, as well as directing doctoral dissertations. Benz has coauthored the recently published book *Getting Started in Writing the Behavioral Dissertation* with Isadore Newman, David Weis, and Keith McNeil. Among her publications are those in *American Educational Research Journal, Review of Higher Education, Journal of Educational Research,* and *Journal of Teacher Education.*